AF673143

FROM SPACE TO SEA

Praise for the Book

'*From Space to Sea* is a valuable contribution to the rather sparse literature on the Indian space programme. The book, written in lucid and easy to read style, is an authentic narration by an expert, who can rightly be called the "father of space propulsion" in India, on diverse chemical propulsion system development ranging from thrust as low as 1 Newton to as high as 4,800 kN of ISRO's launch vehicle programme. The wide canvas of design, development, industry collaboration, collaboration with foreign countries, technology assimilation and absorption, testing, facilities and infrastructure as well as managerial issues are well covered by the author with intimate knowledge of the subject and his deep engagement in the systems realization. Naming and crediting key system contributors is a rare, but welcome feature of the book.'

– **Rajaram Nagappa**, hon. visiting professor, National Institute of Advanced Studies, Bengaluru

'Dr Abraham Muthunayagam, one of the early warriors of the Indian space programme, spent the last few years of his professional career in the development of ocean technologies for national development. He not only created several research and development institutions in this area but also seeded the low temperature thermal desalination programme that could play a major role in addressing the drinking water needs of island communities.

The present monograph is an important and authentic addition to existing literature on India's foray into space and ocean technologies for national development.'

– **Prof. V.S. Ramamurthy**, emeritus professor and former director, National Institute of Advanced Studies and former secretary, Department of Science and Technology

'Filling a huge void in published literature on the Indian space programme, this remarkable book is written by the "father of space propulsion" in India, with interesting technical details. It is also about some internecine conflicts within ISRO. Collaborations with France, Germany, and Russia are described in detail including those connected with Cryogenic engines.

The author is not just an achiever in Space but also in Oceans, another strategic sector, in his later career, which is given in some detail. An unusual but commendable contribution of the author is in naming many unknown engineers and technicians and attributing due credit to them all through the book.

From Space to Sea is a treasure trove for researchers, historians, journalists and writers as well as general readers interested in science and technology developments in independent India.'

– **Y.S. Rajan**, scientist, technologist, writer

'This book is a valuable resource for those who wish to understand the Indian space programme's spectacular growth from launching small imported rockets to becoming a leading space power in its own right. Its author, recruited from the United States by Vikram Sarabhai himself, played a vital part in this progress right from the early days of the fledgling space programme and led the development of liquid propulsion that is needed for launch vehicles as well as satellites.'

– **Gopal Raj,** science journalist and author of *Reach for the Stars: The Evolution of India's Rocket Programme*

FROM SPACE TO SEA

My ISRO Journey and Beyond

ABRAHAM E. MUTHUNAYAGAM

HarperCollins *Publishers* India

First published in India by HarperCollins *Publishers* 2022
4th Floor, Tower A, Building No. 10, Phase II, DLF Cyber City,
Gurugram, Haryana – 122002
www.harpercollins.co.in

2 4 6 8 10 9 7 5 3 1

Copyright © Abraham E. Muthunayagam 2022

P-ISBN: 978-93-5489-888-4
E-ISBN: 978-93-5489-793-1

The views and opinions expressed in this book are the author's own.
The facts are as reported by him and the publishers
are not in any way liable for the same.

Abraham E. Muthunayagam asserts the moral right
to be identified as the author of this work.

All rights reserved. No part of this publication may be reproduced,
stored in a retrieval system, or transmitted, in any form or by any means,
electronic, mechanical, photocopying, recording or otherwise,
without the prior permission of the publishers.

Typeset in 11.5/15.2 Sabon LT Std at
Manipal Technologies Limited, Manipal

Printed and bound at
Thomson Press (India) Ltd

This book is produced from independently certified FSC® paper
to ensure responsible forest management.

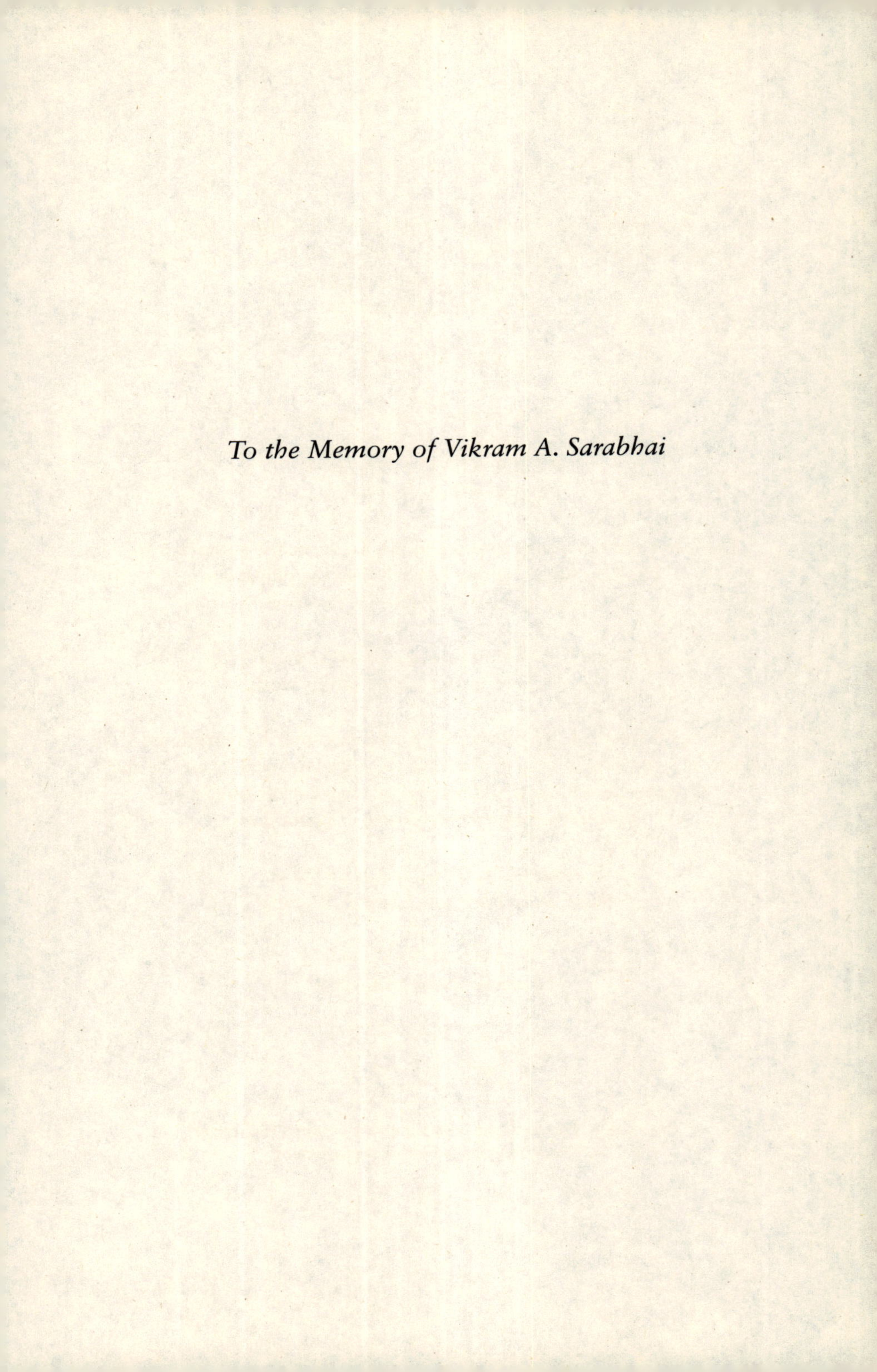

To the Memory of Vikram A. Sarabhai

Contents

List of Abbreviations

ABM	Apogee Boost Motor
ABSD	Ablative Systems Division
ACCE&T	Alagappa Chettiar College of Engineering and Technology
ACOSTI	Atal Centre for Ocean Science and Technology for Islands
ADE	Aeronautical Development Establishment
ADM	Administrative Division
AEDC	Arnold Engineering Development Center
AER	Antarctic Expedition and Research
AIIMS	All India Institute of Medical Sciences
ANCOD	Andaman and Nicobar Centre for Ocean Development
ANCOST	Andaman and Nicobar Centre for Ocean Science and Technology
A&NSFDRA	Andaman and Nicobar Shrimp Farming Development and Research Agency

AOCS	Attitude and Orbit Control System
AOL	Asiatic Oxygen Ltd.
APPLE	Ariane Passenger Payload Experiment Satellite
APSU	Auxiliary Propulsion System Unit
ARD	Aerodynamics Division
ARDE	Armament Research and Development Establishment
ARMEX	Arabian Sea Monsoon Experiment
ASC	Antarctic Study Centre (later renamed NCAOR and then NCPOR)
ASL	Andhra Sugars Ltd
ASLV	Augmented Satellite Launch Vehicle
ASU	Air Separation Unit
ATCM	Antarctic Treaty Consultative Meeting
ATS	Antarctic Treaty System
AutoSal	Skalar Auto Analyser for Nutrient Analysis
AWS	Automatic Weather Stations
BARC	Bhabha Atomic Research Centre
BATL	BrahMos Aerospace Trivandrum Ltd
BEL	Bharat Electronics Ltd
BHEL	Bharat Heavy Electricals Ltd
BIC	Bilateral and International Cooperation
BOBMEX	Bay of Bengal Monsoon Experiment
BSFA	Border Security Force Academy
CCAMLR	Commission for the Conservation of Antarctic Marine Living Resources
CCMHSS	Caldwell Centenary Memorial Higher Secondary School

CD	Convergent-Divergent
CDRI	Central Drug Research Institute
CDS	Catch Documentation Scheme
CECRI	Central Electro Chemical Research Institute
CEP	Committee on Environmental Protection
CESS	Centre for Earth Science Studies
CFD	Computational Fluid Dynamics
CGI	Control, Guidance and Instrumentation Division
CIFT	Central Institute of Fisheries Technology
CISF	Central Industrial Security Force
CIOB	Central Indian Ocean Basin
CLCS	(UN) Commission on the Limits of the Continental Shelf
CMC	Christian Medical College
CMD	Chairman and Managing Director
CMERI	Central Mechanical Engineering Research Institute
CMFRI	Central Marine Fisheries Research Institute
CMLRE	Centre for Marine Living Resources and Ecology
CMM	Coordinate Measuring Machine
CNC	Computerized Numerical Control
CNES	Centre Nationale d'Etudes Spatiales (France)
CNPq	Conselho Nacional de Desenvolvimento Científico e Tecnológico, (Brazil)
CoE	Centre of Excellence
COMAPS	Coastal Ocean Monitoring and Prediction System

COMNAP	Council of Managers of National Antarctic Programmes
CoS	Committee of Secretaries
CPP	Components Production Project
CRV	Coastal Research Vessel
CS	Cryogenic Stage
CSC	Control Systems and Components
CSD	Computer Services Division
CSIR	Council of Scientific and Industrial Research
CSM	Command System Module
CSP	Cryogenic System Project
CTA	Command for Aerospace Technology
CTD	Conductivity-Temperature-Depth
CTPB	Carboxyl-Terminated Polybutadiene
CTRTC	Central Tool Room and Training Centre
CUS	Cryogenic Upper Stage
CUSAT	Cochin University of Science and Technology
CVE	Cambridge Vacuum Engineering
DAE	Department of Atomic Energy
DAT	Disaster Alert Transmitter
DCW	Dhrangadra Chemical Works
DFVLR	Deutsches Zentrum fur Luft-und Raumfahrt
DG	Director General
DGH	Directorate General Of Hydrocarbons
DOD	Department of Ocean Development (now renamed MoEF)

DOS	Department of Space
DPE	Design Project for Electronics
DPGP	Design Project Guidance Package
DPLV	Design Project for Launch Vehicles
DPS	Design Project Stage
DRDL	Defence Research and Development Laboratory
DRDO	Defence Research and Development Organization
DST	Department of Science and Technology
EBW	Electron Beam Welding
ED	Extractive Distillation
EDBOE	Experimental Design Bureau of Oceanological Engineering
EDM	Electrical Discharge Machine
EEZ	Exclusive Economic Zone
EFC	Expenditure Finance Committee
EIA	Environment Impact Assessment
ELD	Electronics Division
ELDO	European Launcher Development Organization
EP	Electric Propulsion
EPDM	Ethylene Propylene Diene Monomer
EPR	Ethylene Propylene Rubber
EPS	Electric Propulsion System
ER&DC	Electronics Research and Development Centre
ERDL	Explosives Research and Development Laboratory

ERP	Expert Review Panel
ERS	European Remote-sensing Satellite
ESA	European Space Agency
ETA	Explosive Transfer Assembly
FAC	Failure Analysis Committee
FCD	Fluid Components Division
FCG	Fluid Components Group
FEP	Fluorinated Ethylene Propylene
FIKEX	First Indian Antarctic Krill Expedition
FLSC	Flexible Linear-Shaped Charge
FM	Frequency Modulation
FNC	Flex Nozzle Control
FORV	Fisheries Oceanographic Research Vessel
FRP	Fibre-Reinforced Plastic
FSI	Fishery Survey of India
FTC	Fin Tip Control
GCRMN	Global Coral Reef Monitoring Network
GK	Glavkosmos
GMDSS	Global Maritime Distress and Safety System
GNSS	Global Navigation Satellite System
GOCO	Government Owned Company Operated
GoI	Government of India
GOOS	Global Ocean Observing System
GPLE	General Purpose Liquid Engine
GPS	Global Positioning System
GSI	Geological Survey of India
GSLV	Geosynchronous Satellite Launch Vehicle

GSLV-Mk-III	Geosynchronous Satellite Launch Vehicle Mark III
GSO	Geosynchronous Orbit
GTO	Geostationary Transfer Orbit
HAL	Hindustan Aeronautics Ltd
HAST	High-Altitude Simulation Test
HAT	High Altitude Test
HE	Head End
HES	Head End Segment
HF	High Frequency
HILS	Hardware-in-Loop Simulation
HOCL	Hindustan Organic Chemicals Ltd
HOD	Head of the Department
HPM	High Performance Motor
HRCSM	Hot Gas Roll Control System Module
HTPB	Hydroxyl-Terminated Polybutadiene
HZL	Hindustan Zinc Limited
ICAR	Indian Council of Agricultural Research
ICMAM	Integrated Coastal and Marine Area Management (later renamed NCCR)
ICZM	Integrated Coastal Zone Management
IDS	Integrated Desalination System
IFFNGPS	Integrated Fish Finder and Navigational Global Positioning System
IFP	Integrated Fisheries Project
IGBP	International Geosphere-Biosphere Programme
IISc	Indian Institute of Science

IIT	Indian Institute of Technology
IKS	Das Interne Kontrollsystem
ILHP	Integrated Liquid Hydrogen Production
ILTP	International Long-Term Programme
IMD	India Meteorological Department
IMO	International Maritime Organisation
INCOIS	Indian National Centre for Ocean Information Services
INCOSPAR	Indian National Committee for Space Research
INSAT	Indian National Satellite System
IOC	Intergovernmental Oceanographic Commission
IOCINDIO	Intergovernmental Oceanographic Commission for the Central Indian Ocean
IOGOOS	Indian Ocean Global Ocean Observation System
IPC	In Plane Correction
IPG	Igniter and Pyrotechnic Group
IPRC	ISRO Propulsion Complex
IR	Inhibition Resin
IRBM	Intermediate-Range Ballistic Missile
IRCGHS	Indo-Russian Centre on Gas Hydrate Studies
IRNSS	Indian Regional Navigation Satellite System
IRS	Indian Remote Sensing
ISA	International Seabed Authority
ISAC	ISRO Satellite Centre (renamed U.R. Rao Satellite Centre)

ISM	Indian School of Mines
Isp	Specific Impulse
ISRO	Indian Space Research Organization
ISS	International Space Station
ISSP	ISRO Satellite System Project
IST	Indian Standard Time
ISTRAC	ISRO Telemetry Tracking and Command Network
ISV	Injectant Start Valve
IUS	Inertial Upper Stage
IWC	International Whaling Commission
JGOFS	Joint Global Ocean Flux Study
KCP	Kheda Communications Project
KELTEC	Kerala Hitech Industries Limited
LAM	Liquid Apogee Motor
LCSD	Launch Vehicle Control System and Components Division
LEO	Low Earth Orbit
LH2	Liquid Hydrogen
LMC	Low Modulus Compound
LMRE	Living Marine Resources Enhancement
LOX	Liquid Oxygen
LPD	Low Pressure Distillation
LPP	Liquid Propulsion Project
LPS	Liquid Propulsion Stage
LPSC	Liquid Propulsion Systems Centre
LPSU	Liquid Propulsion Systems Unit
LPTF	Liquid Propulsion Test Facilities

LSRV	Latchable Series Redundant Valve
LTE	Lakshmi Technology and Engineering Industries
LTTD	Low Temperature Thermal Desalination
MARSIS	Marine Satellite Information Service
MBB	Messerschmitt-Bölkow-Blohm
MCF	Mission Control Facility
MDC	Mild Detonating Chord
MEC	Mechanical Engineering Division
MECZ	Marine Environment and Coastal Zone
MEMS	Micro-Electro-Mechanical System
MEOP	Maximum Expected Operating Pressure
MEV	Main Engine Valve
MIB	Minimum Impulse Bit
MIDHANI	Mishra Dhatu Nigam Ltd
MIP	Moon Impact Probe
MLR	Marine Living Resources
MNLR	Marine Non-Living Resources (programme)
MoEF	Ministry of Environment and Forests
MoES	Ministry of Earth Sciences (formerly DOD)
MOM	Modular Ocean Model
MON-3	Mixed Oxides of Nitrogen
MOU	Memorandum of Understanding
MPD	Magneto Plasma Dynamic (thruster)
MPEDA	Marine Products Export Development Authority
MPN	Multiple Plankton Net
MQC	Materials and Quality Control (Division)

MRCB	Marine Research and Capacity Building
MSI	Maritime Safety Information
MS	Middle Segment
MSMR	Multi-Frequency Scanning Microwave Radiometer
NAL	National Aeronautical Laboratory
NCAOR	National Centre for Antarctic and Ocean Research (formerly known as ASC; later renamed NCPOR)
NCCR	National Centre for Coastal Research (formerly ICMAM)
NCPOR	National Centre for Polar and Ocean Research (formerly ASC and NCAOR)
NDBP	National Data Buoy Programme
NDMA	Nitrosodimethylamine
NDT	Non-destructive Testing
NE	Nozzle End
NES	Nozzle End Segment
NGRI	National Geophysical Research Institute
NHO	Naval Hydrographic Office
NIO	National Institute of Oceanography
NIOT	National Institute of Ocean Technology
NIR	Near-Infra-red
NML	National Metallurgical Laboratory
NPOL	Naval Physical and Oceanographic Laboratory
NPS	Nozzle Protection System
NRSA	National Remote Sensing Agency
NSDRC	National Ship Design and Research Centre

NSTL	Naval Science and Technological Laboratory
NTR	National Test Range
OASTC	Ocean and Atmospheric Science and Technology Cell
OCM	Ocean Colour Monitor
OIS	Ocean Information Services
OOI	Ocean Observation and Information
OOIS	Ocean Observation and Information Services
OOS	Ocean Observing Systems
OPC	Out of Plane Correction
ORC	Overall Review Committee
ORV	Oceanographic Research Vessel
OSTC	Ocean Science and Technology Cell
OTEC	Ocean Thermal Energy Conversion
OV	Orbital Vehicle
PAS	Propellant Acquisition System
PBAN	Polybutadiene Acrylonitrile
PCM	Pulse Code Modulation
PCS	Propulsion Control Systems
PEC	Planning, Evaluation and Coordination
PED	Propellant Engineering Division
PETN	Pentaerithritol Tetranitrate
PFZ	Potential Fishing Zone
PGA	Personnel and General Administration
PMD	Propellant Management Device
PMN	Polymetallic Nodules

PNDF	PSLV Nozzle Development Facility
PPT	Pulse Plasma Thruster
PRL	Physical Research Laboratory
PSLV	Polar Satellite Launch Vehicle
PSM	Pressurization System Module
PSN	Propulsion Engineering Division
PTU	Pressure Transducer Unit
PSOM	PSLV Strap-on Motor
PVC	Polyvinyl Chloride
QC/DC	Quick Connect and Disconnect
R&D	Research and Development
RAS	Russian Academy of Sciences
RCS	Reaction Control System
RFF	Rocket Fabrication Facility
RFNA	Red Fuming Nitric Acid
RH	Rohini
RH-MSR	Rohini Multistage Rocket
RIME	Research Institute of Mechanical Engineering
RLVTD	Reusable Launch Vehicle Technology Demonstrator
RMSA	Remote-mounted Safe Arm
ROCASIN	Rocket Case Insulation
ROV	Remotely Operated Vehicle
RP	Refined Petroleum-1
RPP	Rocket Propellant Plant
RRL	Regional Research Laboratory
RSR	Rohini Sounding Rocket

RTD	Real-Time Decision
RTG	Radioisotope Thermoelectric Generator
S&T	Science and Technology
SAC	Space Application Centre
SAMEER	Society for Applied Microwave Electronics Engineering and Research
SANAE	South African National Antarctic Expedition
SARAL	Satellite for Argos and ALtiK
SCALOP	Standing Committee on Antarctic Logistics and Operations
SCAR	Scientific Committee on Antarctic Research
SCI	Shipping Corporation of India
SCL	Semiconductor Laboratory
SCOUT	Solid-Controlled Orbital Utility Test system
SCRE	Space Capsule Recovery Experiment
SDA	System Development Agency
SDSC	Satish Dhawan Space Centre
SED	Systems Engineering Division
SELMAM	Sea Level Monitoring And Modelling
SEP	Société Européenne de Propulsion
SERC	Science and Engineering Research Council
SFI	Sea Fisheries Institute
SHAR	Sriharikota Range
SIFL	Steel & Industrial Forgings Ltd
SITVC	Secondary Injection Thrust Vector Control
SLB	Solid Rocket Booster
SLV-3	Satellite Launch Vehicle-3
SMP	Solid Motors Project

SPCC	Spacecraft Propulsion, Control Systems and Components
SPM	Special Purpose Motor
SPROB	Solid Propellant Space Booster Plant
SPSD	Satellite Propulsion Systems Division
SPT	Stationary Plasma Thruster
SR	System Reliability
SRB	Solid Rocket Booster
SRM	Solid Rocket Motor
SROSS	Stretched Rohini Satellite Systems
SSC	Space Scientific Committee
SSD	Satellite System Division
SSO	Sun-Synchronous Orbit
SSPL	Solid State Physics Laboratory
SST	Sea Surface Temperature
SSTC	Space Science and Technology Centre
SSV	System Start Valve
STEX	Static Test and Evaluation Complex
STR	Structural Engineering Division
TBI	Through Bulkhead Initiator
TCFC	Technical Coordination and Finance Committee
TDV	Technology Demonstrator Vehicle
TERLS	Thumba Equatorial Rocket Launching Station
TES	Technology Experiment Satellite
TP	Turbopump
TPS	Thermal Protection System
TVC	Thrust Vector Control

UDMH	Unsymmetrical Dimethylhydrazine
UK	United Kingdom
UKWP	UST-Katav, Wagon Building Plant
UNCED	United Nations Conference on Environment and Development
UNCLOS	United Nations Convention on the Law of the Sea
UNCLOS III	Third United Nations Conference on the Law of the Sea
UNEP	United Nations Environment Programme
UNESCO	United Nations Educational, Scientific and Cultural Organization
URSC	U.R. Rao Satellite Centre (formerly the ISRO Satellite Centre)
VASIMR	Variable Spacecraft Impulse Magneto Plasma Rocket
VCCG	Vehicle Control Systems and Components Group
VHF	Very High Frequency
VIP	Very Important Person
VMC	Vessel Management Cell
VSSC	Vikram Sarabhai Space Centre
VTM	Velocity Trimming Module
WG-EMM	Working Group on Ecosystem Monitoring and Management
WMO	World Meteorological Organization
XBT	Expendable Bathythermograph
YOTO	(International) Year of the Ocean

List of Photos

Preface

SPACE FLIGHT IS A significant and remarkable human achievement of the twentieth century. Tentative attempts to fly to outer space began in the 1950s and grew explosively in subsequent decades. The multifarious applications of space technology have now impacted all walks of life – be it satellite TV reaching even the remotest villages, tribal settlements and islands through rooftop dishes, daily weather maps, cyclone warnings, the ubiquitous Global Positioning System (GPS), Google maps or fresh knowledge about the universe through interplanetary missions. Not to mention the development of military systems with advanced surveillance capabilities, missiles and other powerful force multipliers.

India has emerged as a spacefaring nation – joining the ranks of six others on our planet – the United States of America (USA), the erstwhile Union of Soviet Socialist Republics (USSR), China, France, Germany and Japan – and has made tremendous progress in communication, earth observation, navigation and scientific infrastructure in space, including the recent explorations of Moon and Mars.

Propulsion systems play an inevitable and important role in all these ventures. Our scientists have mastered propulsion technologies, used for Indian Space Research Organization

(ISRO) launch vehicles and satellites programmes. The operational launch vehicles of ISRO – PSLV (Polar Satellite Launch Vehicle), GSLV (Geosynchronous Satellite Launch Vehicle) and GSLV-Mk-III (Geosynchronous Satellite Launch Vehicle Mark III) – have made significant contributions to space initiatives and are acknowledged worldwide.

Nevertheless, very little is available in the public domain about this great saga of pioneering efforts – the technical challenges, international collaborations, technology denials and bureaucratic red tape, to mention a few strengths and weaknesses. This book is primarily an attempt to tell the story of my career over four decades in two hitherto unexplored realms – space and the ocean. I look back with pride and satisfaction on my association with ISRO and Dr Vikram Sarabhai (Photo No. 1), who chose to give me the responsibility, in 1965, of initiating the development of propulsion systems in India to meet the space mission's requirement of launch vehicles and satellites. Simultaneously, this book endeavours to enlighten the general reader on the intense drive of the Indian scientific mind partaking in the most advanced technology and applying the knowledge gained to render the life of the common man more secure and comfortable.

In the course of my career at ISRO, I had to face several challenges, most of which I converted into opportunities and successfully accomplished organizational goals with the sincere and dedicated contributions of my colleagues. The tasks were often multidisciplinary and could be tackled successfully only by teamwork and selfless cooperation. However, even as I enjoyed a professionally rewarding career, there was considerable intra-organizational politics that led to the denial of promotions I was eligible for on time and recognition at the national level. When matters seemed to reach a stalemate, I requested to be shifted from ISRO and was appointed as secretary to the Department of Ocean Development (DOD), Government of India.

After serving national space and ocean programmes for about thirty-five years, I shifted my attention to application-oriented programmes that would benefit the community, in association with academic institutions and industries. Recognizing the threat underlying the fast-depleting sources of potable water in the country, I focused my efforts on creating a sustainable, economical and reliable desalination system, using renewable solar energy. I envisioned that it would be a joint venture between a consortium of Indian industries, academia and the Makai Ocean Engineering Company from the US, with the requisite technology transfer from the US company. Such innovative ventures can succeed only with the support of our government.

Since I am still working towards this goal and am yet to establish an operational project, this book does not delve into my preoccupation with and work in this area over the last two decades. Nor does it go into my post-retirement engagements – as the director and founder vice chancellor of Karunya University, as the executive vice president of the Kerala State Council for Science, Technology and Environment, my part-time association with IIT Madras as its chairman of the board of governors and as pro chancellor of the Noorul Islam Centre for Higher Education in Kanyakumari.

Since my superannuation, I have been sorting out personal affairs and spending valuable time with my family and, overall, leading a life of contentment. However, some of my friends and former colleagues requested me to put down my professional experiences in writing to create a comprehensive historical record and also to inspire future generations in space research and development. Aware of the inherent merits of such a project, I have taken up this work at an advanced age with the support of and inputs from my contemporaries and well-wishers.

Organization of the Book

This book has twenty-two chapters. It begins with an outline of my early life in a laid-back village in Tamil Nadu, then traces my long journey from prestigious institutions of higher learning in India and abroad to the ups and downs of a coveted career at ISRO, and finally with a brief account of my path-breaking assignments in interdisciplinary areas. The first chapter titled 'Early Life' is a nostalgic account of my childhood, an exemplary education and an extremely fortunate selection to the innovative Indian Space Programme. Chapter two describes the growth of ISRO from a historical perspective and my association with the great visionary Dr Vikram A. Sarabhai. The next chapter further elucidates the pioneering efforts in space during the Vikram Sarabhai era. As the title indicates, chapter four 'The Satish Dhawan and Brahm Prakash Era' explores the progress in space propulsion at the Vikram Sarabhai Space Centre (VSSC) under stalwarts such as Dr Satish Dhawan and Dr Brahm Prakash.

The subsequent chapter familiarizes the readers with Vikas, the ISRO launch vehicle workhorse engine, and productive foreign collaborations such as the Viking engine technology transfer from Société Européenne de Propulsion (SEP), France. Chapter six deals with the production, supply and technology developments at the Pressure Transducer Unit (PTU) in Bengaluru. The chapter that follows this examines the realm of solid propulsion systems, while chapter eight delineates my endeavours as the programme director of the Auxiliary Propulsion System Unit (APSU) in the monopropellant and bipropellant Reaction Control System (RCS) development for the launch vehicles of ISRO.

Chapter nine commences with an account of the emergence of satellite propulsion systems in India and then talks about the monopropellant propulsion for Indian Remote Sensing (IRS) and bipropellant propulsion for the Indian National Satellite System (INSAT) series of satellites, as well as the technology transfer for titanium tanks from Germany. The ensuing two chapters venture into the realm of cryogenic propulsion systems and the collaboration with Russia, and offer an in-depth exposition on the components of 'propulsion and control systems' for the IRS and INSAT series, the launch vehicles SLV-3 (Satellite Launch Vehicle-3), ASLV and PSLV, and liquid propulsion stages.

Chapter twelve provides valuable first-hand knowledge of ISRO's French, German and Russian collaborations on propulsion test facilities. Following this is a chapter that describes Mahendragiri, ISRO's world-class propulsion test centre, set up on the Mahedragiri Hills in Tamil Nadu. The subsequent chapter dedicates itself to the first three launch vehicles of ISRO: SLV-3, ASLV, PSLV. The fifteenth chapter reveals the personal frustrations that accompanied my professional achievements and my consequent bowing out of ISRO, as I moved to the Department of Ocean Development (DOD).

The final chapters, 16 to 22, are exhaustive accounts of the radical transformation of DOD with national programmes on Antarctic expeditions and polar science, Marine Non-

Living Resources (MNLR), Marine Living Resources (MLR), Ocean Observation and Information Services (OOIS), Integrated Coastal and Marine Area Management (ICMAM), ocean technologies, capacity building, and international programmes, United Nations Convention on Law of the Sea (UNCLOS), and Inter-governmental Oceanographic Commission (IOC) and also the research ventures I initiated in DOD's nationwide associated units, with the cooperation of the political leadership and specialists in various disciplines.

The appendix addresses the ever-expanding boundaries of space propulsion, gives a brief history of rocket propulsion and the classification of propulsion technologies, and points to emerging trends in the field for those keenly interested in the discipline. I hope this book appeals both to the curious reader as well as the specialist, and motivates students to pursue research that benefits the nation.

Prologue

THE MAIDEN FLIGHT OF the Liquid Apogee Motor (LAM) on 11 July 1992 still remains fresh in my memory. I was at the Mission Control Facility (MCF) in Hassan with the ISRO team to witness the launch. The first manoeuvre to raise the INSAT-2A spacecraft launched by Ariane-4 from the Geosynchronous Transfer Orbit (GTO) was scheduled for 6.10 a.m. IST (Indian Standard Time). INSAT-2A, weighing 1,906 kg, was built indigenously, and its LAM was developed by the Liquid Propulsion Systems Centre (LPSC).

All eyes at the MCF were glued to the monitors that displayed the propulsion parameters. The normally closed pyro valves and the latch valve in the gas module were commanded to open as planned. On account of excessive leakage through the pressure regulator in the system, the pressure in the propellant tank steadily increased and reached 22 bar, nearing the burst pressure of the tank, which was designed for an operating pressure of 16.3 bar.

The scientists present grew tense as they recognized the need for spontaneous decisions and quick action at that moment. Thankfully, ISRO's wisdom, knowledge and experience prevailed, in addition to divine intervention. Loud and almost impulsive actions taken by everyone around forced the LAM to be fired immediately before reaching the apogee, approximately 200 seconds ahead of the estimated time. A catastrophe was avoided.

The LAM started firing, and the pressure in the propellant tank started coming down to the nominal, with the propulsion system now performing satisfactorily. Relief was evident on all the faces at the MCF after 3,932 seconds, which was 80 per cent of the planned total burn. The Geosynchronous Orbit (GSO) was attained with one more burn of the LAM in blowdown mode for 806 seconds. The LAM had triumphantly demonstrated its capabilities in extreme operating conditions. It is heartening to note that this LAM has successfully performed over fifty ISRO missions without a single failure.

1

Early Life: From the East Coast to the West

THE INCREDIBLE JOURNEY OF my career began over five decades ago. On 15 April 1966, I first set my eyes on the Thumba Equatorial Rocket Launching Station (TERLS) on the outskirts of Thiruvananthapuram. It marked the start of my involvement in the Indian Space Programme, which ultimately spanned three decades. Groups of scientists and engineers that I established there have shaped and continue to lead the development, testing, production and assembly of liquid propulsion systems essential for India's launch vehicles and satellites.

During my tenure, ISRO progressed from launching small, imported rockets, known as sounding rockets, used to study the upper atmosphere of the earth, to building bigger and more powerful ones that could put satellites into orbit. Experimental satellites paved the way for the country to produce its own images of the earth, keep watch on its weather, and provide communication and broadcasting assistance.

This phase of my life was full of technological challenges and opportunities, which culminated in my appointment as secretary to the Department of Ocean Development, Government of India. There, I addressed Antarctic research programmes, the United Nations Convention on the Law of the Sea, marine living and non-living resources, the Ocean Observation and Information Services (OOIS), Integrated Coastal and Marine Area Management (ICMAM) and the Intergovernmental Oceanographic Commission (IOC) of the United Nations Educational, Scientific and Cultural Organization (UNESCO), among others. But all of that came much later. My beginnings were rather humble.

My Family

My parents – Muthia Abraham and Chinnammal Abraham – were schoolteachers. Apart from discharging the normal parental duties such as providing me with food, clothes and shelter, they trained me in Christian ethics, instilled in me the need to love God and our fellow men, and taught me human values that turned out to be the guiding principles in my life.

I am the eldest son of my parents. I had a younger brother – Dr A.H. Diaraj, a geologist from Madras University. He specialized in geophysics at Indiana University and served in major oil and gas companies in America. Unfortunately, he passed away at a young age in the 1990s.

During my postgraduate days at IISc, in May 1961, I got married to Ms Ida William, the second daughter of Dr C. William of Nagercoil. A graduate in physics from Scott

Christian College in Nagercoil, she preferred to look after the family rather than pursue higher studies or take up a job. The attention she gave to our kids during their childhood was laudable. Since our children got married and settled away from Thiruvananthapuram, we have been living alone, with a full-time domestic worker to help us. We lead a Christian life blessed with joy and peace.

We have three sons. The eldest is Dr Abraham W. Muthunayagam, a urologist who runs the Muthunayagam Hospital in Nagercoil. He completed his undergraduate, postgraduate and super-speciality medical studies at the Christian Medical College (CMC), Vellore. He is married to Renu S. Das, daughter of Dr Sugunan K. Das from Thiruvananthapuram. They have two children, Dr Benjamin Abraham and Dr Benita Abraham. Our second son is Dr Jonathan V. Muthunayagam, a cardiologist from CMC Vellore, who shares a group practice with his in-laws at Salem, Tamil Nadu. He married Dr Aruna T. Kiruba, the daughter of Dr C. T. Kiruba from Salem; they also have two children – Dr Cheruba Jonathan and Dr Roshan Jonathan. Our youngest son, Dr Newton P. Muthunayagam, is a gastroenterologist in Cleveland, USA. After his DM in gastroenterology from Madras Medical College, he interned for three years at Amarillo, Texas, and availed himself of the two-year fellowship programme at the Cincinnati Medical School before moving to Cleveland. His wife is Dr Neena M. Bennet. Her father superannuated as director of Medical Services in the Government of Tamil Nadu. They are blessed with two daughters – Cynthia Newton and Ria Newton – who attend schools in Cleveland.

Idaiyangudi, My Village

I was born on 11 January 1939 at Idaiyangudi, a small village of about 150 predominantly Christian families in Tirunelveli district of Tamil Nadu. Idaiyangudi, once an impoverished region along the east coast of India, benefited greatly from the missionary work of the Reverend Robert Caldwell from Ireland. He came to Idaiyangudi in December 1841 and developed the area in a phased manner by establishing a school, a hospital and a church. Until then, in the absence of proper education and healthcare, people earned a living from doing odd jobs, climbing palm trees to collect toddy, selling fish from the Uvari coast and engaging in agricultural activities. Modernity had not set its foot in the locality.

There were no medical facilities with qualified doctors in rural areas for decades after I was born, but Idaiyangudi had a Christian Mission Hospital, administered by the Church of South India, with the infrastructure to admit patients for treatment. There was no high school in the neighbourhood. So the older children had to walk nearly five miles every day to school and back. In the 1950s, the school – now the Caldwell Centenary Memorial Higher Secondary School (CCMHSS) – had a total strength of more than 1,000 students, of which over 80 per cent were from outside my village. I successfully completed the SSLC Board examination in 1954, with a first class and as the school topper.

As a boy, I used to play football in the school playground. We also had a ball badminton court within our compound at home, where friends would join us siblings for a game. Other hobbies included gardening and reading. There were no recreational facilities such as a cinema, zoo or museum nearby.

Puttu (rice cake) with sugar and plantains constituted my breakfast on most days; rice with fish curry/fry comprised lunch and dinner. We had a fish market where fresh fish from the coast was available daily around 11 a.m. The adjoining fishing village Uvari, situated along the coast of the Bay of Bengal, was nearly a mile away. Uvari boasted of a prominent Catholic church and a Hindu temple; devotees from distant places visited the latter during the annual festival. I frequented the Uvari beach with family and friends to watch the never-ending waves washing the shores and to play at the sand dunes. Occasionally, we revelled in the shallow waters and followed it up with a freshwater bath from the nearby wells.

After my studies at CCMHSS, I joined the two-year intermediate course of Madras University at St John's College, Palayamcottah, in Tirunelveli district. I graduated in 1956 with a first class, after which I applied for admission to the engineering degree course at the Alagappa Chettiar College of Engineering and Technology (ACCE&T), Karaikudi. I was awarded a free-tuition scholarship for the entire course. During those days, I used to play football and volleyball, and represented the college in inter-college volleyball tournaments. In 1960, I graduated in mechanical engineering with honours, securing a first class and the second rank in Madras University.

I continued my education with a master's programme in power engineering (mechanical) at the Indian Institute of Science (IISc) in Bangalore (now Bengaluru) and graduated in 1962 as the topper in the discipline.

My Desire for Higher Studies Abroad

During my postgraduate studies at IISc, I was attracted to the courses on heat transfer and thermodynamics, both of which were taught by Dr Arcot Ramachandran, the head of the department (HOD). Impressed by his teaching methodology, administrative skills and cordial interactions with students, I made him my role model. My desire then was to take up research, adopt teaching as a profession, work in an academic institution in Tamil Nadu and simultaneously focus on application-oriented projects for the socio-economic benefit of the nation. To achieve these objectives, it was essential to have a doctoral degree, preferably from abroad. I planned to go for a doctoral degree programme in heat transfer, thermodynamics and allied areas, if possible.

I learnt from my batchmates that financial assistance was available in US universities, in the form of paid, part-time teaching or research assistantship for those who undertook postgraduate studies. I approached Dr A. Ramachandran, expressed my desire for higher education to him, and requested him for guidance and help to get an admission to any US university that offered such financial support. He gave me the names and addresses of three people who were working in heat transfer and allied areas: Professors Yeram S. Touloukian at Purdue University, Indiana, Jim Hartnett at Delaware University and John Clark at the University of Michigan. I was glad to get favourable responses from all three institutions, with offers for admission and part-time research assistantship to cover my expenses overseas.

I chose Purdue University, where Dr Ramachandran had done his doctoral work in the area of heat transfer. After the successful completion of my postgraduate degree course at IISc, I joined Purdue in September 1962 for the doctoral degree programme. Dr W. Leidenfrost was my major professor at the Mechanical Engineering School. I also availed myself of a part-time research assistantship at the university.

Life at Purdue

During the first (Fall) semester, I stayed at the married students' quarters of the university with my Indian friends – Kasim Sinnamohideen and K. Nagappan – from Tamil Nadu. Our residence was barely a kilometre away from the lecture halls of the university.

Over the course of my first winter in 1962, I experienced some difficulty in walking through the frozen snow to school and back in order to attend the morning lectures. I also had trouble adjusting to American food. However, I got used to the system soon and began to prepare my own meals with rice, meat, fish, eggs, etc. Initially, there were issues in communication due to the unfamiliar accent of my fellow American students.

Near the departments at Purdue, there were houses with independent rooms and good facilities, including common cooking areas. I rented one of these rooms with good amenities for the 1963 winter semester and continued there for the rest of my studies. I prepared Indian food, and watched TV in the common room. Every evening, after dinner, I used to return to the department to work on the experiments for my doctoral degree and to spend time at the university library. I worked very hard and obtained an A grade in all the subjects, for all the semesters except the first. I successfully defended my thesis, fulfilled all the requirements of the university and received a PhD by April 1965 within thirty-two months of joining the university. It was an achievement to be proud of.

I had been employed for about a year at the FluiDyne Engineering Corporation, an aerospace company in Minneapolis, when I received an offer from the Indian National Committee for Space Research (INCOSPAR) to join the Indian Space Programme at Thumba, Thiruvananthapuram.

While I was studying at Purdue University, my wife and our eldest son had remained at her parental home in Nagercoil. I chose to join my family in India, so I returned during the first week of April 1966 and then shifted to Nanthancode in Thiruvananthapuram to take up the job at Thumba. My parents lived with us during the initial days and guided our household routine. I started my work at ISRO on 15 April 1966. It was the Tamil New Year's Day, an auspicious beginning to an exceptionally rewarding career.

From Purdue to ISRO

Inspired by the vocation of my parents, I had gone to the US to qualify to become a great teacher in a higher institute of technology in India, like my role model Dr A. Ramachandran of IISc. A simple advertisement in *India News* in 1964 unexpectedly changed the direction of my life.

During my college days, I was interested in learning about research programmes in the science and technology ministries of our country, especially in the fields of atomic energy, space and defence. I fervently wished to contribute to national progress in an academic institution of repute, in the specific area of my expertise. I learnt about the successful launch of Nike-Apache, a two-stage US rocket from Thumba on 21 November 1963, marking the beginning of the Indian Space

Programme. In the process of the launch, TERLS became operational under INCOSPAR, which had been established in 1962, with Dr Vikram Sarabhai as the Chairman. Little did I know then that I would soon be associated with the programme.

In 1964, during my studies at Purdue University, I came across an advertisement of INCOSPAR in *India News*, a periodical the university received from the Indian embassy. The advertisement mentioned the beginning of an 'Indian Space Programme' and invited applications from Indian nationals keen to work in different disciplines in space science and technology in the country. I sent in my application. Later, while I was working at FluiDyne Engineering Corporation, INCOSPAR contacted me for an interview with Dr Vikram Sarabhai at the Indian embassy in Washington DC. In 1965, I had a brief interview with Dr Sarabhai. Dr Upendra Desai of the Physical Research Laboratory (PRL), then on a deputation to the Goddard Space Flight Center of NASA, assisted Dr Sarabhai in the selection process.

On the date of the interview, all the candidates were waiting in front of the embassy. Dr Sarabhai walked towards us and introduced himself. No Indian official accompanied him. His simplicity and pleasant manners impressed everyone. In the course of the interview, Dr Sarabhai briefly summarized his plans and the challenges that lay ahead for the nation. He further asked each one of us about how we could contribute to the space initiative. The same afternoon, around 4 p.m., Dr Sarabhai met all the candidates shortlisted, held further discussions with each of them individually, finalized his selection and offered those chosen various positions at Thumba.

A group of six was recruited from the US: Dr S.C. Gupta, Dr D. S. Rane, Dr Y.J. Rao, Dr M.K. Mukherjee, M.C. Mathur and myself. It must be noted that this interview was not against any specified number of posts or grades in salary. I was the youngest of the lot with the lowest pay, but the remuneration was normalized at the end of the first year of employment.

Since I had a pre-existing commitment to a NASA project at FluiDyne, I was granted permission to join INCOSPAR after its completion.

Around the same time, Dr Sarabhai recruited R.M. Vasagam, K. Kathirvel, R. Vivekanandan, S.N. Prakash and Dr C.V. Ouseph in India. Subsequently, Dr V.P. Kulkarni from the USSR, Dr C.L. Amba Rao from the US and Dr V.R. Gowariker from the United Kingdom (UK) also joined our team. The professionals handpicked by Dr Sarabhai from all over the world possessed the knowledge, skills and expertise to evolve and implement the Science and Technology (S&T) projects relevant to the Indian Space Programme. We were the pioneers appointed at the Space Science and Technology Centre (SSTC) in 1967 under the guidance of Dr Sarabhai, who initiated the space research programme at Thumba. We converted the challenges in the formative stages of the Indian space venture into excellent opportunities with the support of Dr Sarabhai and the ISRO management.

I reported for duty in April 1966 to H.G.S. Murthy, Director, TERLS, at Thumba and started my work at ISRO in the area of propulsion. Thus, at twenty-seven years of age, I moved from the east coast to the west coast and eased into the beginning of my professional stint in India. Little did I know then that the bulk of my career would be in this capital city of Kerala on the coast of the Arabian Sea.

Dr Sarabhai formulated the Indian space policy, which was definitive for years to come and would inspire all of us: 'There are some who question the relevance of space activities in a developing nation. To us there is no ambiguity of purpose. We do not have the fantasy of competing with economically advanced nations in the exploration of the Moon or the stars or manned space-flights. But we are convinced that if we are to play a meaningful role nationally, and in the community of nations, we must be second to none in the application of advanced technologies to the real problems of man and society.'*

His striking personality influenced many of us to work with him in the Indian Space Programme for decades, to envisage and participate in the phenomenal growth of the organization and its consistent contribution to national development.

---

* Quoted in 'List of Important Speeches and Papers by Dr Vikram A. Sarabhai', https://prl.res.in/library/sarabhai_v_speeches.pdf

2

Journeying with Dr Sarabhai: Technology Centre, Thiruvananthapuram

INDIA EMBARKED ON THE launch vehicle development programme with the SLV-3 in 1972 and the Augmented Satellite Launch Vehicle (ASLV) in 1982. Today, ISRO has operational launch vehicles such as the PSLV, GSLV and GSLV-Mk-III – all of which have made significant contributions to Indian space missions. Simultaneously, ISRO launched the Aryabhata, Bhaskara and Rohini series of satellites. At present, the organization has the IRS series for remote sensing, the INSAT series for communication and the Indian Regional Navigation Satellite System (IRNSS) series for navigation.

Over the years, India has made tremendous progress towards enhancing payload launch capabilities and developing satellites for application in communications, earth observation, navigation, space exploration and scientific infrastructure, to name a few areas. In all these endeavours, propulsion systems play an inevitable and challenging role. ISRO has also lined up exciting future missions with great challenges – the Small Satellite Launch Vehicle, the Reusable Launch Vehicle and the All Electric Propulsion Satellite, among others.

ISRO and the Sarabhai Era

The successful launch of Nike-Apache, a two-stage US rocket, from Thumba on 21 November 1963 marked the beginning of the Indian Space Programme, with TERLS becoming operational under INCOSPAR on the same day. In the initial stages, the research and development (R&D) activities began at Thumba with the offices functioning inside the St Mary Magdalena Church (Photo No. 2) and the nearby Bishop's House.

Dr Vikram Sarabhai had an exceptional vision for the Indian Space Programme, which was translated into missions and executed by the eminent leaders of ISRO. He believed in sustainable development with indigenous technology and acquiring know-how from developed nations, when required. With his vast experience in both the government and private sectors, Dr Sarabhai adopted an innovative approach to organize new activities, establish institutions and identify executives to lead challenging tasks. He recognized the multidisciplinary nature of, and the skills required for, the space programme and took personal interest in the area of human resource development. He put in special effort to induct experts working abroad in space science and technology and allied areas into the Indian Space Programme. He approached Indian nationals with the requisite knowledge, expertise and potential through Indian embassies and his contacts overseas.

Space Science and Technology Centre (SSTC)

Dr Sarabhai integrated all the R&D activities of INCOSPAR at Thumba by establishing the SSTC in 1967. He recognized the importance of leadership in organizing and managing such an important national institution and took over the responsibility as the founder director of the SSTC. He organized different divisions at the SSTC with the infrastructure and manpower necessary to nurture expertise and to establish technologies that met the challenging needs in the wide spectrum of areas in space science and technology.

Dr Sarabhai initially set up the SSTC with the following divisions: Propulsion Engineering Division (PSN), Propellant Engineering Division (PED), Aerodynamics Division (ARD), Structural Engineering Division (STR), Materials and Quality Control Division (MQC), Mechanical Engineering Division (MEC), Control, Guidance and Instrumentation Division (CGI), Electronics Division (ELD), Systems Engineering Division (SED), Computer Services Division (CSD) and the Administrative Division (ADM). He appointed divisional heads with specific responsibilities and the authority to execute tasks in an effective and efficient manner. I was designated the divisional head for the PSN and the MEC in 1967. Thus, I started my career with the duty of developing solid and liquid propulsion systems at INCOSPAR, which was renamed ISRO in August 1969. Dr Sarabhai simultaneously established multidisciplinary and multi-divisional projects with definite objectives and budgets with individual project executives.

Technical Coordination and Finance Committee (TCFC)

Dr Sarabhai adopted an unconventional management system for the SSTC. He appointed a five-member Technical Coordination and Finance Committee (TCFC) to assist him. It comprised Drs V.R. Gowariker, S.C. Gupta, M.K. Mukherjee, Y.J. Rao and myself. The chairman of the TCFC would represent the director of the SSTC during the latter's absence at Thumba. Each member would hold the chair of the TCFC for a period of three months on a rotational basis. The approving authority for all proposals submitted by the chairman of the TCFC would be the member who succeeded him at the end of the three months' tenure. In addition to the regular responsibilities in their respective divisions, these TCFC members were nominated for additional assignments at the SSTC as well.

Dr M.K. Mukherjee was in charge of the SSTC administration and recruitment. Dr S.C. Gupta saw to the construction of new buildings and the establishment of infrastructure and service facilities. I was the head of financial matters, Dr Gowariker interfaced with national chemical research institutions, and Dr Y.J. Rao interacted with national aerospace institutions. It is interesting to recall the manner in which Dr Sarabhai administered the SSTC as its director. He visited the premises once a month, inspected the laboratories, held discussions with engineers and other officials, and reviewed its progress. He conducted the TCFC meetings on the SSTC campus and, at times, over dinner. The meetings would continue until all the items on the agenda were dealt with.

The TCFC met at least once a week, normally on Wednesdays during lunch, and continued discussions until the completion of its agenda. The deliberations consisted of status and progress reviews, approvals for expenditures, proposals for initiating activities, new facilities and infrastructure, and the creation of additional posts and recruitment, among other things. The collective wisdom of specialists in multifarious disciplines in the day-to-day management of the SSTC was effective. The act of working together and resolving complex issues by taking timely decisions promoted the concept of teamwork and collective leadership. It was a great experience, despite the occasional differences of opinion among the members. I interacted frequently with Dr Sarabhai and S.R. Thakore of the PRL who assisted him in all financial matters.

Dr Sarabhai – the Ultimate Inspiration

I had an excellent understanding and working relationship with Dr Sarabhai, both professionally and personally. This book on my journey with the organization would be incomplete without the enriching details of my association with this great man. He was my first boss in India. I worked directly under him for about six out of my twenty-nine years at ISRO. I can proudly say that Dr Sarabhai has always been the guiding force in my professional life.

Dr Sarabhai was a great achiever who set goals with focus and established several institutions of relevance to social welfare and national development. He spearheaded institutions in the space and atomic energy sectors of our nation. He demonstrated that innovation in management and capable leadership in execution were essential for success. Dr Sarabhai had a unique vision for the Indian Space Programme.

He familiarized himself with the evolution of space ventures in the developed nations, and maintained personal relationship with key personnel in the space programmes in the US, France and Japan. He wanted to gain from the experience of others and to evolve initiatives that met our national needs; he chose not to reinvent the wheel.

Dr Sarabhai did not believe in the exclusive ownership of territories or areas in Science and Technology (S&T) research. He sought international cooperation in the arenas of science and technology wherever possible in order to catch up with more accomplished countries. He encouraged constructive competition among colleagues as it brought out creativity, encouraged hard work and achieved quick, efficient results. Dr Sarabhai gave ample opportunities for youngsters to carry out innovative research relevant to the ISRO programmes. He gave them freedom and found time to directly interact with them.

Dr Sarabhai was a man of quick action. He guided and helped his subordinates accomplish tasks on time. In this context, I would like to narrate an incident that is etched in my memory. Dr Sarabhai wanted me to attend a meeting in the US at short notice, within two days. In those days, there were restrictions on foreign travel. Reserve Bank of India's permission, foreign exchange and international travel tickets from Air India were mandatory. However, on Dr Sarabhai's instructions, I travelled to Bombay (now Mumbai) and stayed at the PRL guest house on Pedder Road (now renamed Gopalrao Deshmukh Marg). The next morning I visited the Air India office, the Reserve Bank and the American Express office, all located in the Flora Fountain area. To my amazement, all arrangements had already been made by Dr Sarabhai's office. I collected my approvals, air tickets, a traveller's cheque and so on, and left for New York the same night.

Above all, Dr Sarabhai was a great human being. He was simple and never publicly displayed his authority as the secretary to the Government of India. He was very cordial, a patient listener, extremely considerate and fair in decision making. He maintained personal connections with his colleagues who were key personnel in various units. Dr Sarabhai frequently held informal meetings with them. Such sessions took place during breakfast, while travelling in cars or even in flights. At times, he asked colleagues to accompany him on flights, where he would use the time to discuss issues, take decisions and issue directions.

I still remember a TCFC meeting held in a car driven by Dr Y. Janardhana Rao from the Trivandrum (now Thiruvananthapuram) airport to the hotel at Kovalam where Dr Sarabhai stayed. His usual driver had been excused as the deliberations were on confidential matters. The Ambassador car was full with Dr Sarabhai and the five TCFC members.

Our relationship was not between a boss and his subordinates. He never issued orders; instead, he requested his subordinates to get things done on time. All these qualities added to his achievements.

Waiting for a Breakfast Meeting

I would like to conclude this chapter with the sad end of my association with Dr Sarabhai. On 29 December 1971, he visited the SSTC as usual. That night I was invited for a discussion with him at his hotel in Kovalam. We had dinner in the company of N.V.G. Warrier, his private secretary. On 30 December, we were scheduled to have a TCFC dinner meeting at the same hotel. We talked about the industries emerging around the Thumba complex. When we dispersed, he asked me to join him for breakfast the next day. Sadly, this breakfast meeting never took place.

As I was preparing to go to Kovalam, I was informed of his sudden demise and was asked to travel to Ahmedabad with his body along with my TCFC colleagues and others. I lost my boss, my mentor and a good friend forever, an irreparable vacuum in my life.

Among many treasures, the greatest lesson I learnt from him is: 'Shift your wisdom in words to wisdom in action. Be bold enough to set goals, believe that you can, and be brave enough to do everything possible with dedication and commitment.'

Throughout my professional life I have tried to follow these guidelines to the best of my ability.

3

Rocket Propulsion at the SSTC

DURING THE FIRST DECADE of its operation, TERLS mostly launched foreign rockets, which included the following:

(i) American: Nike-Apache, Nike Tomahawk, Arcas and Judi-Dart

(ii) French: Centaure and Dragon

(iii) British: Skua-I, Skua-II and Petrel, and

(iv) Soviet Union: M-100.

Soon, however, efforts were made to produce sounding rockets in India. In 1964, Dr Sarabhai concluded an agreement with Sud Aviation, a French aerospace company, to manufacture under licence a two-stage sounding rocket named Centaure. He then established the Rocket Propellant Plant (RPP) at Thumba in 1968 to produce solid propellant blocks, under licence from France, and a Rocket Fabrication Facility (RFF) in 1969 to manufacture the hardware for Centaure.

At the same time, work on the indigenous sounding rockets called Rohini also began. The experience gained through the RPP and RFF in the production-under-licence programme significantly contributed to the development of indigenous sounding rockets, with improved performance, in the shortest time possible. It must also be noted that the Rohini sounding rockets programme laid the foundation for launch vehicle development at ISRO.

Indigenous Propulsion System Development for Sounding Rockets

Since solid propellants were still in the formative stages at the SSTC, the Propulsion Engineering Division (PSN) had to select one from external sources to carry on with the solid propulsion technology development. The Cordite Defence Factory at Aravankaud in the Nilgiris, produced double-based extruded solid propellants in two different sizes for their projects. Along with Dr Y.J. Rao and T.S. Ram, I visited the factory and held discussions with the officials regarding the feasibility of using those cordite propellant blocks in developing the solid rocket motors.

Based on the specifications provided by the Cordite Defence Factory, I designed a 75mm diameter solid propellant motor. Dr Y.J. Rao was the project leader for Rohini-75 (RH-75); his team carried out the flight analysis, and designed the nose cone, base shroud and fins for this motor. We designed the first Indian solid rocket to be accepted by Dr Sarabhai. Following due approval from the defence authorities, the

Cordite Defence Factory supplied the available double-base extruded propellant blocks for the initial solid propulsion system development to PSN.

I actively participated in the design and development of the first indigenous 75 mm diameter rocket motor RH-75, which was successfully flight-tested in November 1967. It was the first effectively flight-tested small rocket made at ISRO, Thumba. My responsibilities included grain design, interior ballistic analysis, charging and assembly of the solid motor, ignition system development and the establishment of the assembly and test facilities. Until the commissioning of the RPP in 1968 and the successful development and supply of indigenous composite propellants, the double-base propellants from the Cordite Defence Factory were used in rocket motors.

The main objective in the development of RH-75 was to get hands-on experience in designing and developing different elements, assembling and integrating the rocket motor system and evaluating the performance of the propulsion system on the ground and flight tests. In these efforts, we used commercially available extruded aluminium alloy tubes for the motor case, commercial resins for charging the grain, insulating rubber shoes at the head end and the nozzle end of the rocket motor for thermal protection.

During the development of RH-75, my PSN team experienced some schedule criticality in producing the insulating rubber shoes at the Trivandrum Rubber Works. As an alternative, I procured commercially available liquid rubber from Chennai. My team, under the leadership of T.S. Ram, had worked out the composition and process for the production of rubber shoes for insulation in the RH-75 rocket motor by adding different quantities of inert solid fillers, plasticizers, curatives and so on to the liquid rubber. Subsequently, the requisite insulating rubber shoes for RH-75 were made by following this process, using the commercial liquid rubber in specially prepared moulds.

After the successful launch of RH-75, I proposed to replace the inert solid fillers used in the production of the insulating rubber shoes with ammonium perchlorate, the oxidizer used in the propellant, and attempted to produce a composite propellant with this liquid rubber as the binder. Dr Sarabhai approved this change. When an oxidizer is substituted in the place of an inert material, we get a propellant, a combustible product, instead of insulating material. Thus this step was a redundant 'propellant development activity' at the SSTC, in addition to the ongoing developments carried on by PED.

Having successfully developed propellants for the 50 mm diameter zero-zero motor, attempts to flight-qualify the propellant in RH-75 continued by both the PSN and PED teams: with the polyvinyl chloride (PVC) binder at PED and the rubber binder at PSN. After three successive successful ground tests, both propellants qualified for flight. On 21 February 1969, 'Mrinal', the propellant developed by PED, was successfully tested in the dynamic test vehicle RH-75. The next day, the propellant developed by my PSN team was also successfully flight-tested in RH-75. Thus, Dr Sarabhai's encouragement of healthy competition paid rich dividends and made two composite propellants available in a short time.

It is noteworthy that the redundant propellant development by my team in the PSN served as a catalyst to propellant developments in PED. My team discontinued their propellant development efforts after the successful flight test, but PED carried on their work with different binders and compositions, and developed several higher-energy propellants. Consequently, in January 2010, we saw ISRO's largest and the world's third largest solid propellant motor, with about 200 tonnes of state-of-the-art, fully indigenous, solid propellant, clear a static test at the Satish Dhawan Space Centre (SDSC), Sriharikota. Two such huge (3.2 m diameter) motors were used in the 610-tonne GSLV-Mk-III, the biggest launch vehicle of ISRO, which was successfully flight-tested in May 2017.

Operational Sounding Rockets: RH-200, RH-300 and RH-560

At the PSN, I established the technical capability, infrastructure and manpower required to design, develop and supply solid propulsion systems for all the sounding rockets of ISRO. Under my supervision, solid rocket motors for projects RH-100, RH-125, RH-200, RH-300 and RH-560 were developed and supplied. I also established facilities for rocket assembly, propellant charging and static testing for small-diameter solid motors at Thumba. During the 1967–75 period, a series of Rohini rockets with different capabilities was designed and developed. It resulted in the current Rohini Sounding Rocket (RSR) family, including RH-200, RH-300 and RH-560.

As chairman of the Rohini Sounding Rockets Review Committee, constituted by ISRO Chairman Dr Satish Dhawan, I reviewed the status of developments in ISRO sounding rockets and brought out a detailed report in 1977. It addressed the status, design adequacy and problem areas, and offered some recommendations. This report was accepted and implemented.

Judi-Dart Rocket-Import Substitution

TERLS launched the Judi-Dart rocket supplied by NASA in the 1960s to study coastal winds at lower altitudes. It had a 75 mm diameter rocket motor named 'Judi', carrying a 25 mm diameter payload 'Dart', packed with a copper chaff. This chaff was released at around 50 km altitude. The copper chaff got dispersed and drifted along with the wind. The ground-based radar tracked the chaff to study the atmospheric wind pattern.

H.G.S. Murthy, Director, TERLS, initiated the development of an Indian version of Judi-Dart under the leadership of A.P.J. Abdul Kalam, who was in charge of range safety and payload integration activities at TERLS. The Dart payload was successfully developed by Abdul Kalam. He designed a 75 mm diameter motor to replace Judi. Since no solid propellant was available at Thumba, he approached the Explosives Research and Development Laboratory (ERDL) in Pune, which supplied a composite propellant. He carried out the static tests of the motor at Pune and the flight test at TERLS. Unfortunately, Kalam could not deliver the chaff payload at the specified 50 km altitude. Further flight tests were abandoned since the rocket did not have the capability required to meet the specified objective.

Subsequently, to establish the indigenous capability of the imported Judi-Dart rocket, two versions of RH-100 were taken up for development – the first one using the RPP composite propellant and the motor designed and developed by my PSN team, and the second one using the PED composite propellant and the motor designed and developed by Abdul Kalam's team. After successful static tests, both versions were flight-tested at TERLS. The RH-100 rocket with the motor developed by my team successfully deployed the copper chaff at 50 km and met the set objective. Consequently, ISRO gave up the development of the RH-100 piloted by Abdul Kalam's team. However, since the required information on coastal winds at 50 km altitude became subsequently available from other sources, this project was ultimately discontinued.

Multi-Stage Rocket: Project RH-MSR

To promote technology developments in different aspects of the space programme, Dr Sarabhai established multidisciplinary and multi-divisional projects with particular goals, budgets and project executives. These ventures were in addition to promoting research, development and human resources, and establishing infrastructure in the various SSTC divisions. In 1968, Dr Sarabhai established the Rohini Multistage Rocket (RH-MSR), a multistage rocket project, at the SSTC. Concurrent to my assignments as the head of the PSN and MEC divisions, he appointed me as the project director so that I could gain experience in the staging and clustering of rocket motors with the small-diameter rockets available.

As the project leader, I developed a 125 mm diameter rocket motor using indigenous composite propellant produced at the RPP. This motor was flight-tested successfully as a single-stage rocket (RH-125), a two-stage rocket with RH-125 as the booster and RH-75 as the sustainer and a two-stage rocket with three RH-125s (clustered with a common manifold at the head end) as the booster and a fourth as the sustainer, all in 1968. Thus, certain technologies related to the staging and clustering of rockets were studied and demonstrated in-flight in a primitive way, using the small-diameter rockets available.

In addition, the cluster block motor technology to develop high-thrust, short-duration-burning solid motors, using multiple free-standing grains, was also demonstrated. A cluster-grain motor, with the cordite propellant, was flown as a booster for the first 600 kgf thrust liquid engine as sustainer in 1973. This project gave limited hands-on experience in the design, analysis, ground-testing and flight-testing of different configurations, with the rocket motors available. No operational configuration with these small-diameter rockets was foreseen for any application. Hence the project was short-closed and the resources were diverted towards SLV-3.

The major contribution of this project was the production and supply of more than a hundred RH-125 rockets with the propellants for diverse applications. Towards the end of 1970, as directed by Dr Sarabhai, I transported six RH-125 motors and the dummy nose cones to the Border Security Force Academy (BSFA) at Tekanpur, Gwalior, Madhya Pradesh. I demonstrated the integration of the dummy nose cone to the RH-125 motors and flight-tested all six of them with the assistance of the BSF team. The BSF jawans at Tekanpur were thrilled to launch RH-125 rockets for the first time.

In 1971, I supplied quite a few RH-125 rockets without the nose cone to the BSF. Major Bhatnagar, the BSF coordinator, informed me that he procured the shells of 125 mm diameter from the Armament Research and Development Establishment (ARDE) in Pune, assembled them to the RH-125 motor and flight-tested these rockets for demonstration and training.

4

The Satish Dhawan and Brahm Prakash Era: Space Propulsion at the VSSC

AFTER DR VIKRAM SARABHAI'S demise, all space activities at Thumba were brought together under the Vikram Sarabhai Space Centre (VSSC) in 1972. This centre was given the responsibility of developing the technology needed for the ISRO's launch vehicle programme. Dr Brahm Prakash, who enjoyed a distinguished career in the Department of Atomic Energy (DAE), became the first director of the new centre after his superannuation. He was chosen by Dr Satish Dhawan, Chairman, ISRO, to reorient and restructure the R&D programmes of the VSSC based on his excellent performance at the Bhabha Atomic Research Centre (BARC) of the DAE in the field of metallurgy. He reorganized the VSSC along the lines of BARC with groups, divisions and facilities with group directors, divisional heads and general managers.

Dr Brahm Prakash was a mature, senior professional with considerable expertise and years of managerial experience. He always had a smiling face and never got angry or shouted at anyone. He patiently listened to the views of others, was not biased and took decisions based on the merit of the case. He consulted all the people concerned, along with his staff officers – Dr S. Ramnath and R.M. Manickavasagam – and examined new schemes/proposals in depth. When he found the need to disagree with a proposal, he discussed it with the proposer and convinced the person of the rationale behind his decision. He had an excellent understanding of and working relationship with Dr Satish Dhawan. He contributed significantly to the restructuring of the VSSC and evolved the management plan for the first satellite launch vehicle SLV-3. It was a pleasure to work with him.

Dr Brahm Prakash made me the group director of the propulsion group responsible for developing propulsion systems for the launch vehicles and satellites of ISRO. From 1972, I evolved, guided, reviewed and approved the activities of the VSSC propulsion group with my colleagues at the time: Head of Solid Propulsion Rajaram Nagappa, Head of Liquid Propulsion S. Nambinarayanan, Head of Control Systems and Components Sridharan Dhas, Head of Heat Transfer and Thermodynamics Dr B.C. Pillai, Head of Liquid Propellants (Oxidizers) R. Natarajan, and Head of Liquid Propellants (Fuels) V. Udayanathan.

Initial In-house Liquid Propulsion Systems Development

ISRO recognized the significant role of liquid propulsion systems in launch vehicles with multiple applications.

However, during the 1970s there was no approved project for any major liquid propulsion system development except the bipropellant and monopropellant Reaction Control Systems (RCS) for the control of the second and third stages of SLV-3. This was due to pre-existing organizational commitments to the then ongoing SLV-3 and sounding rocket projects, and the financial constraints caused by them.

During the late 1960s, when we were engaged in the development of sounding rocket and planning for the first launch vehicle that relied on solid propellants for all four solid propulsion stages, I felt the need to initiate some in-house R&D in the area of liquid propulsion. I approached Dr Sarabhai then and, subsequently, Dr Satish Dhawan in 1972, expressing my desire to develop liquid propulsion systems in a small way in order to gain the knowledge and experience essential for meeting ISRO's needs in future. Dr Sarabhai appreciated the proactive proposal and approved our scheme, which had a minimal budget so as to not adversely affect the other commitments.

I started work by selecting commercially available raw materials and making use of in-house fabrication facilities and human resources to create the requisite hardware and establish assembly and test facilities. Initially, I decided to work on earth-storable, hypergolic propellant systems and chose aniline as fuel and RFNA (Red Fuming Nitric Acid) as the oxidizer. We experimented with different injection systems with impinging jets and swirl jets. We selected the pressure-fed system for propellant supply and film-cooled ablative liners for thrust chamber cooling. We developed all the essential control components in-house – the solenoid valve, pressure regulator, check valve and filters and so on.

1. Dr Vikram Ambalal Sarabhai, father of Indian Space Programme, is the founder director of Space Science & Technology Centre at Thumba and the author's first boss in his professional career in India

2. St. Mary Magadelene church at Thumba where the first office was established for space engineers and scientists in 1966

3. Static test of 200 kgf thrust bip-ropellant engine at the horizontal test bed at Thumba beach

4. Static test of 600 kgf thrust bi-propellant engine at the vertical test bed at Thumba beach

5. Static test of liquid stage with 3000 kgf thrust bi-prppellant engine stage at the engine-cum-stage test bed at Shriharikotta

We had designed and fabricated the first bipropellant liquid engine with a thrust of around 5 kgf in our workshop and tried it in a test bed with minimum facilities at a remote corner in Veli Hills in SSTC in 1969. All of us were happy that Dr Sarabhai witnessed one of the tests at Veli. A. Chandran, our engineer, and Dhanabalan, a technician, were key persons in the endeavour. Subsequently, we shifted our test operations to the newly constructed test bay at Thumba beach, near the Centaure solid motor test stand. We designed, manufactured and built a horizontal test stand to test liquid thrusters with minimum support facilities for assembly, integration, instrumentation and propellant service. We developed solenoid valves and designed a 10 kg thrust hypergolic bipropellant engine with on–off capabilities by using aniline and RFNA propellants. We manufactured the hardware in our workshop and demonstrated the engine in the test bay. Thus the feasibility for developing an on–off RCS for launch vehicle control was demonstrated.

With the technology and facilities available, and under the leadership of K. Sivaramakrishnan Nair, we established a standard demonstration set-up for visitors in the block house at Thumba beach with the 10 kg thruster, a control panel and pen recorders. VIP (Very Important Person) visitors are first shown our thruster – assembled on the test stand, charged with the propellants and nitrogen gas, and ready for test. They are then taken to the control room. After explaining the operations and demonstrating a few of the thruster's on–off pulses, they are given the control switch and requested to operate the thruster at random mode from the control panel. Whenever the thruster is turned on, the loud noise of its firing is heard and the chamber pressure of the thruster is simultaneously recorded by the pen recorder in the control panel. The guests are naturally thrilled to witness the thruster responding to their commands and to see the recorded pressure of the thrust chamber. It is a fascinating experience.

Simultaneously, we started the development of monopropellant thrusters with hydrazine as well as hydrogen peroxide as propellants. After some progress, we decided to continue with the development of monopropellant systems using hydrazine alone. The major challenges were the development of an iridium-based catalyst and the injector head. A small quantity of imported 'shell' catalyst was experimented with. A.K. Ray and C.G. Parameswaran, along with K. Rajasehara Rao from the material division of the VSSC, worked round the clock with the testing team to achieve this. In the testing area, R.V. Raman and Arunachalam Pillai were preoccupied with the instrumentation and M.R. Prasad, R. Gilbert, Lukose, C.K.K. Nair and others were engaged in the assembly of hardware and in the preparation of the thrusters. In 1977, two monopropellant RCS packages were designed, realized and flight-tested to gain expertise in system development. The first one was the de-spinning of the Centaure rocket, using 1 kgf monopropellant thrusters; the second was for the attitude control of RH-560, using 1 kgf and 25 kgf thrusters.

In 1981, another RCS package had to be developed for the Ariane Passenger Payload Experiment Satellite (APPLE) for its attitude control with sixteen 1 Newton hydrazine monopropellant thrusters. The indigenous catalyst development could not meet the schedule. Hence, it was decided to import the system from Hamilton Standard in the US, along with the propellant servicing cart to save time. This step gave everyone involved valuable experience and inputs for the later satellite RCS development programmes of ISRO.

At that time, the focus of the PSN was on bipropellant thrusters. Subsequently, we developed a 200 kgf thrust, ablative-cooled liquid engine, (Photo No. 3) which we tested in the horizontal test stand. Maintaining the specified mixture ratio between oxidizer and fuel was a requirement for guaranteed performance. During the development trials, we finalized a procedure to be followed for water calibration and decided to introduce orifices in the fluid circuits for flow control and obtaining the mixture ratio required.

After several successful tests, we established a vertical test stand, next to the horizontal test stand, to test higher-thrust engines with all their accessories. Then we developed a 600-kgf thrust liquid engine with the same hypergolic liquid propellants, ablative-cooled thrust chamber and axial-impinging multiple injectors, burning for about twenty seconds. The continuous roar of the burning 600 kg engine on the shores of Thumba, accompanied by long vertical flames, was very impressive (Photo No. 4). It became another attraction for visitors.

We successfully tested the 600 kgf engine several times. During one of his visits to Thumba in the early 1970s, Dr Dhawan was impressed by the engine test and suggested that we configure a liquid vehicle with this engine and launch it from Sriharikota to gain hands-on experience in assembling and flying liquid rockets. Towards flying the first liquid rocket of ISRO, we configured the liquid stage with two propellant tanks (one for fuel and the other for oxidizer), inter-stage structure, base shroud with fins, nose cone, etc.

We fabricated all hardware in-house and found that the 600 kgf engine could not provide the required take-off acceleration for the stable flight of the liquid rocket when it was fully loaded with the propellant. Hence, an additional booster stage was required to fly this liquid propellant rocket. For this purpose, we made use of a short-burning, clustered cordite propellant solid booster, which was developed under the Rohini Multistage Rocket (RH-MSR) project. We also designed and fabricated a launcher in-house to launch this two-stage liquid rocket with a solid booster.

We transported the rocket, launcher, propellants, hardware, etc., to SHAR (the Sriharikota Range renamed as the Satish Dhawan Space Centre [SDSC] in 2002) and started our launch campaign in 1973. During the integration of the vehicle with rigid pipes connecting the propellant tanks and the engine, we experienced some difficulties in achieving the required vehicle alignment with leak-proof joints. We were able to identify the need for flexible hoses for easy assembly and leak-proof joints, which we introduced in later programmes. We established the launcher, propellant storage and transfer facilities and other support facilities in one of the SHAR launch pads (Photo No. 6).

We carried out two launches of the 600 kgf liquid vehicle from SHAR between 1973 and 1974. Owing to financial constraints, the flight configuration was simple without any on-board instrumentation. On the first flight, the liquid stage was ignited in-flight with a pyrotechnic delay. On the second flight, we incorporated a vented inter-stage skirt and ignited both the booster and the liquid stage simultaneously on the ground. Since we had no instrumentation on board, we had no record to confirm the ignition of the liquid stage in the first flight. But the second flight was during the night and we could see, with our naked eyes, the trajectory of the liquid stage from the flame discharged.

Thus the developments in liquid propulsion systems during the initial stages were:

- 1969 to 1972: Development of bipropellant engines with 5, 10, 200 and 600 kgf thrusts;
- Joint development, with the materials and quality control division, of a 1 kgf thrust monopropellant engine which focused on the molybdenum-based catalyst; and
- 1973: First liquid rocket flights – 600 kgf thrust engine.

With the experience gained both on the ground and in the flight testing of the 600 kgf thrust liquid engine in simple configuration with minimum expenditure, we started the development of a 3-tonne liquid engine in 1973. When successfully developed and optimized with high-energy liquid propellants and high-performance thrust chamber and injection system, it could yield an upper liquid stage for any one of the later ISRO launch vehicles. But to start with, we used the same hypergolic propellant system with ablative-cooled thrust chamber. We designed, fabricated and established a vertical test stand for the 3-tonne project in the Static Test and Evaluation Complex (STEX) at SHAR during 1976, when both the engine and the integrated stage were tested and qualified (Photo No. 5).

We carried out two flights of this 3-tonne liquid stage with minimum on-board instrumentation. The first one was as the second stage of a two-stage vehicle, with the proven RH-560 as booster, to provide adequate acceleration at take-off; the second was a single stage (Photo No. 7). From the telemetry data of the vehicle, we observed that the liquid stage did not ignite in the first flight due to the absence of a positive propellant discharge device from the tank. In the second flight, the burn time of the liquid stage was less than that predicted, on account of the incomplete discharge of all propellants to the engine in the absence of a proper intake device on the engine intake lines.

While these two flight experiments were not wholly successful, we gained substantial experience in the assembly, integration and launch operations of liquid propulsion stages. In addition, we gained hands-on experience in propellant filling, gas pressurization, operational safety and so on. From the first flight, we learnt the need for detailed analyses and studies to understand the fluid motions in the propellant tank during acceleration and deceleration during flight, and to provide positive expulsion devices in the tank to ensure the required propellant discharge to the thrust chamber.

From observation of the second flight, we recognized the need to understand the fluid motions inside the propellant tank of a spinning vehicle in flight and to provide anti-vortex baffles/devices in the propellant intake to the engine to prevent vortex formation and early gas discharge through the engine before the propellant is fully discharged. Thus, these two flights gave very valuable insights into the design and development of liquid stages for flight. Consequently, we were able to critically address the propellant acquisition systems and anti-vortex baffles for every flight in future. Further, the experience from such preliminary experimentation contributed significantly to liquid propulsion technology transfer programmes from SEP, France, as would be explained later.

Simultaneously, with the ongoing development of the 3-tonne engine, we addressed the development of high-performance propulsion systems for later ISRO programmes. We initiated studies on different types of feed systems, which included but was not limited to stored gas-pressure-fed systems for low thrust and low burning time, blowdown, pressure-fed systems for higher thrust and short duration burning, gas-generator-pressure-fed systems for low thrust and higher burning time and turbopump-fed systems for higher thrust and higher burning time and so on. It was noted that the turbopump (TP) systems with gas generators were commonly used in high-thrust liquid engine developments.

We designed and initiated the development of two gas generators for the 3-tonne project – the first one for the pressure-fed system using a gas generator and the second one to drive the turbopump. In the gas generator design, we replaced the aniline-RFNA propellant system with the hydrazine-RFNA-water system to eliminate soot formation and reduce exhaust temperature. The gas generator for the pressure-fed system was designed for operation at a pressure of 320 bar with a flow rate of 0.15 kg/s and an exhaust temperature of 500°C. The gas generator for the turbopump-fed system was designed for operation at a pressure of 15 bar with a flow rate of 0.34 kg/ and an exhaust temperature of 800°C. We also designed and initiated the development of a turbopump for the 3-tonne project.

Moreover, we established a hydraulics laboratory for the (1) analysis of injector characteristics, which included local mixture ratio, mass flux distribution and spray mean droplet size; (2) flow calibration of injectors, feed lines, flow control components, etc.; and (3) flow visualization through axisymmetric bodies by hydraulic analogy. In addition, we established a gas jet facility for nozzle ablative evaluation to find heat transfer coefficients with pyrolysis and ablation and to develop charring and ablation models. While our initial studies were with aniline and RFNA propellants, we shifted to N_2O_4-hydrazine propellants and carried out studies with carbon phenolics, silica phenolics and asbestos phenolics ablatives.

A summary of the major activities during the 3-tonne liquid stage development is presented below:

- 1974: Engine development with axial injector
- 1974: Pilot plant for propellants
- 1975: Commissioning of hydraulics laboratory
- 1976: Engine development tests
- 1977: Stage development tests
- 1977: Gas generator development
- 1978: Turbopump development, and
- 1978: First flight test of the 3-tonne liquid stage rocket.

Liquid Propellants

Yet another major requirement for the development of a liquid propulsion system is the availability of oxidizers and fuels in large quantities. In the initial stages, we had chosen commercially available, low-energy aniline and RFNA as propellants because they were not expensive. But high-energy liquid propellants were essential to develop high-performance propulsion systems. In the liquid propellant division of my group, I established a team of chemical engineers to carry out laboratory experiments to finalize the process, production and handling methodologies, and to evolve a scheme to make high-performance liquid propellants available in the country in large quantities to cater to the ISRO programmes. We decided to interface with chemical industries in India and chart out a system and procedure to ensure that these propellants were produced and supplied to ISRO.

Towards this goal, I established two project teams for studies and pilot-scale experiments. The first was under the leadership of R. Natarajan, chemical engineer, who was assigned the responsibility for the oxidizer nitrogen tetroxide (N_2O_4) and mixed oxides of nitrogen (MON-3); the second team functioned under the leadership of V. Udayanathan, chemical engineer, with the responsibility for the fuels, unsymmetrical dimethylhydrazine (UDMH) and monomethylhydrazine (MMH). The work started with a 5 kg-level pilot plant for N_2O_4 that employed lead nitrate decomposition route and a 50-kg-level pilot plant for UDMH that used nitrosodimethylamine (NDMA) route. We combined the results of our studies with the experience of Indian industries to arrive at an industrial production programme.

Oxidizers

Nitrogen Tetroxide (N_2O_4)

At that time, Hindustan Organic Chemicals Ltd (HOCL) was in the process of establishing a concentrated nitric acid plant, using a high-pressure route via NO_2, at Rashayani, Mumbai. Our engineers approached HOCL and discussed the results of our pilot project on decomposition of lead nitrate. In consultation with the HOCL experts, we chose their process to establish the production plant for N_2O_4 by condensing

NO_2 to N_2O_4 and purifying the liquid to propellant grade specification. In 1979, ISRO signed an agreement with HOCL and a production unit was commissioned at HOCL within three years. Today, HOCL regularly supplies N_2O_4 to ISRO at the rate of around 600 tonnes per year.

Mixed Oxides of Nitrogen (MON-3)

The Liquid Propulsion Systems Centre (LPSC) developed a batch process to produce MON-3 in the laboratory, using N_2O_4 as the raw material. A plant adopting the batch process with a capacity of 150 kg/batch was established at Mahendragiri in 1990. It still supplies MON-3 regularly and meets all the requirements.

Fuels

Anhydrous Hydrazine (N_2H_4)

This compound is used as fuel in the bipropellant thrusters and as the propellant in the monopropellant thrusters in the control systems of the launch vehicles and satellites of ISRO. Based on the results of a pilot plant study, a production plant to process 100 litres of hydrazine hydrate per batch was established at the VSSC in 1978 to produce anhydrous hydrazine (N_2H_4) from hydrazine hydrate (N_2H_4 H_2O), adopting extractive distillation, using sodium hydroxide (NaOH). This plant meets all the needs of the ISRO programmes, both monopropellant and bipropellant.

Unsymmetrical dimethylhydrazine (UDMH) and Monomethylhydrazine (MMH)

For the production of UDMH, the initial studies that employed nitrosodimethylamine (NDMA) route were abandoned due to the carcinogenic nature of NDMA. The team, under the leadership of V. Udayanathan, a chemical engineer, started work on the chloramine process for both UDMH and MMH synthesis. Yet another process, using unsymmetrical dimethylurea (UDMU) to produce UDMH, was also developed. A national committee under the chairmanship of Dr S. Varadharajan, secretary to the Government of India, was constituted by the ISRO chairman to review and recommend a process for industrial production. The committee suggested pilot plant studies of both processes – the UDMU route at IDPL, Hyderabad, and the chloramine route at the Andhra Sugars Ltd (ASL), Tanuku, West Godavari district, Andhra Pradesh.

Based on the results of the studies, the chloramine process was selected for both UDMH and MMH in 1979. Parallel efforts were also initiated to produce the key raw materials – methylamines – within the country. The Fertilizer Corporation of India Ltd (FCIL) was approached and a production plant was established in their facility at Rashtriya Chemicals and Fertilizers Ltd (RCF), Mumbai. In 1990, at the request of ISRO, ASL established a production plant for UDMH and started supplying it at the rate of 120 tonnes per year. The plant was augmented with an additional distillation unit for MMH production with a capacity of 25 kg per day. Subsequently, another plant of similar capacity was installed for UDMH. Since 1990, ASL has been producing UDMH and MMH and still supplies ISRO regularly.

Cryogenic Propellants

When we initiated the development of cryogenic propulsion systems for GSLV, we required large quantities of liquid oxygen (LOX), liquid hydrogen (LH_2) and liquid nitrogen (LN_2). In 1987, I assigned the responsibility to supply these cryogenic fluids (LOX, LH_2, LN_2) to a project team under the leadership of V. Udayanathan. We explored the possibility of establishing the production plant in one of the existing industries with the raw materials and infrastructure. We approached DCW Ltd (Dhrangadhra Chemical Works Ltd) and other alkali and fertilizer plants located in south India. Unfortunately, we could not identify an Indian manufacturer that could supply cryogenic fluids to our specifications. So I proposed to establish an Integrated Liquid Hydrogen Production (ILHP) facility to produce all these cryogenic at the Liquid Propulsion Test Facilities (LPTF) in Mahendragiri, and to assign the responsibility for operation and maintenance to an industry with experience in producing and handling gaseous hydrogen.

Based on the interactions and discussions with the industries working on hydrogen and other cryogenic fluids, we decided to use naphtha as feedstock to produce gaseous hydrogen and to adopt pre-cooled expansion of liquid nitrogen to produce liquid hydrogen. Thus, the ILHP facility at LPTF, Mahendragiri, included an air separation unit (ASU), gaseous hydrogen (GH_2) synthesis and purification unit with naphtha as feedstock, hydrogen compressor unit, cold box for LH_2, storage and transport tankers, and all other allied utilities. Following due administrative procedure, Linde plc in Munich, Germany, was selected as the prime contractor to establish the ILHP facility to produce and supply the cryogenic fluids with annual production capacities of 150 tonnes of LH_2, 600 tonnes of LOX and 700 tonnes of LN_2. An agreement was signed with Linde plc on 4 February 1991.

On 23 March 1992, we laid the foundation stone for the Integrated Liquid Hydrogen Production (ILHP) facility at Mahendragiri. In 1992–93, equipment and machinery were manufactured and supplied by Linde plc and Gardner Cryogenic, USA. Under the leadership of V. Udayanathan, my team coordinated with these industries abroad and commissioned ILHP facility at Mahendragiri in 1995. The responsibility to operate and maintain the plant and fulfil the cryogenic fluid needs of ISRO was contracted to Asiatic Oxygen Ltd (AOL). Considering the critical nature of and the need for timely availability of the cryogenic fluids, ISRO decided to establish a redundant source of supply. In 2003, a second liquid hydrogen plant with a capacity of 150 tonnes per year was established at ASL for the production of LH_2 by using the gaseous hydrogen available at their alkali plant.

I am very pleased to record that LPSC established production plants both in-house and in industries to produce and supply all propellant requirements for the earth-storable and cryogenic propulsion systems. I congratulate V. Udayanathan, R. Natarajan and their team members for their dedicated and committed efforts towards the success of our liquid propulsion systems development.

5

Liquid Propulsion Systems and Vikas, the Workhorse of ISRO Launch Vehicles

HIGH-PERFORMANCE AND HIGH-THRUST LIQUID propulsion systems were inevitable for the ISRO space programme. Therefore, the need arose to embark on a major liquid propulsion system development programme on a project mode. I recall the in-house activities for the development of liquid propulsion systems initiated in 1969 under major financial constraints and the early developments using commercially available low-energy propellants. The efforts made towards the development of, and the designs for, 3-tonne-thrust-level propulsion subsystems, including gas generators, turbopumps, thrust chambers, etc., and the studies on high-energy propellants and the infrastructure for their development, were also memorable.

Liquid Propulsion Systems

I realized that the experience gained during these developments and the flight testing of the 600-kgf-thrust and 3-tonne-thrust liquid stages were very valuable but not adequate to develop high-performance and high-thrust liquid propulsion systems for ISRO programmes in an accelerated, time-bound manner. In 1980, Europe had established Ariane, a multinational company to develop launch vehicles. The Société Européenne de Propulsion (SEP), an aerospace company in France, was responsible for the development and supply of propulsion systems for the Ariane programme, in association with allied industries in Europe. Hence, I considered Ariane to be a potential technology provider for ISRO in the development of liquid propulsion systems.

Dialogue with the Commercial Director of SEP

During the days when the Static Test and Evaluation Complex (STEX) was designing test facilities for SLV-3 solid propulsion systems jointly with SEP under my responsibility, I had an opportunity to pass on two pressure transducers, developed by R.M. Vasagam of the SSTC and used in the static tests of solid motors to R. Morin, the commercial director of SEP, with a request to evaluate them for our test programmes at STEX. He assessed the two pressure transducers and informed me that ISRO had demonstrated the basic technology for manufacturing pressure transducers but it was necessary to incorporate temperature compensation for multiple uses in aerospace applications. He added that SEP could provide ISRO with the transducer technology, based on which we could manufacture and supply pressure transducers of aerospace quality to allied industries across the globe. He further suggested that SEP establish a collaborative programme with ISRO as 'consultant and technology provider' in our diverse launch vehicle programmes for mutual benefit. He pointed out that SEP was in the final stages of qualification of the 60-tonne thrust-rated (nominal) Viking engine for Ariane. The possibility of an ISRO participation in the programme on mutually beneficial terms existed.

Collaboration with SEP

I realized that ISRO was in the midst of critical developments in SLV-3s, the first all-solid, four-stage vehicles. However, on account of a severe financial crisis, the organization could not divert funds to any new initiative. Nevertheless, I briefed Dr Dhawan, our Chairman, on the opportunity for collaboration with SEP and the technology transfer of the Viking engine (Photo No. 8). He appreciated my views and agreed that ISRO could consider a mutually beneficial collaboration, provided there was no transfer of funds. Further, ISRO had experienced and qualified engineers whose services could be made available for the development programmes of SEP. He added that ISRO had access to manufacturing facilities, which could make and supply products for SEP developments. He was willing to consider the proposed collaboration with SEP, if the services and supplies from ISRO could compensate for the technology transfer.

I conveyed this message to R. Morin, who was keen on continued collaboration between SEP and ISRO, beyond the STEX test facilities. He then discussed the idea with the senior management of his firm, and it was favourably considered. Soufley, the President of SEP, and Dr Dhawan met in India; they discussed the details and agreed in principle to the proposal. Dr Dhawan authorized a two-member team of T.N. Seshan, additional secretary in the Department of Space (DOS), and me as the director of the Propulsion Group (VSSC) to negotiate with the SEP management and submit a mutually acceptable proposal. In pursuance of this idea, the teams from SEP and ISRO discussed the venture in depth and arrived at a satisfactory recommendation. In these meetings, R. Morin represented SEP. T.N. Seshan addressed the administrative and financial aspects, and I dealt with technical and related management duties. The summary of the recommendations made were:

(i) SEP should transfer to ISRO the available technology of the Viking engine and subsequent developments till the third launch of the Ariane II vehicle.

(ii) SEP should transfer to ISRO the available technology for the 21U and 21NS* pressure transducers. It should also provide all the documents required on the designs, manufacturing drawings, assembly and integration, and the calibration procedures of these transducers. In addition, SEP should supply the special purpose Electron Beam Welding (EBW) machine, a vacuum brazing furnace and six precision lathes to manufacture the transducers.

(iii) ISRO was to provide the services of 100 man-years to work on different development activities related to the Ariane vehicle at SEP and their subcontractors. The assignments for 75 per cent of them would be decided by SEP and the remaining 25 per cent was to be at the discretion of ISRO.

---

* 21U and 21NS are designations given by SEP for transducers.

(iv) ISRO should continue to pay these personnel their salaries in India without any change in their service conditions. SEP should provide a subsistence allowance of 2,400 Fernch franc in France for 75 per cent of the people who worked on SEP assignments. ISRO agreed to pay the same subsistence allowance in France for the remaining 25 per cent of those who were engaged in ISRO assignments.

(v) ISRO should manufacture and supply about seven thousand 21U and 21NS pressure transducers according to the requirements of SEP regarding the type, quantity, pressure range, delivery schedule and so on. In turn, SEP should supply all the materials required for the manufacture of these transducers with 10 per cent excess to account for rejections, if any.

(vi) The detailed operational systems and procedures were to be worked out and approved by ISRO and SEP.

An agreement was executed in 1974 by Seshan and Souffley on behalf of their respective organizations. I was present on this momentous occasion, along with Dr Dhawan and other officials of ISRO. In this context, I would like to point out that T.N. Seshan coined the name 'Vikas' for the Viking engine project. Vikas means 'development' in Sanskrit. It is also a word derived from Vikram A. Sarabhai.

The Vikas Project

To implement the ISRO–SEP agreement, the Vikas Project commenced at the VSSC in 1974, and I was the project director. The responsibility was in addition to my other assignments as group director at the VSSC. This collaboration was indeed a major breakthrough in the development of liquid propulsion systems in ISRO. As the project director of Vikas, I organized and administered the ISRO–SEP initiative for over a period of five years.

Deputations to SEP

As per the agreement that had been signed, ISRO had to provide the services of its engineers to SEP for 100 man-years. In consultation with SEP, I finalized the assignment, the period of deputation and the preferred qualification and experience of the personnel to be sent to SEP. In association with the VSSC, I selected suitable candidates from my propulsion group and other divisions of the VSSC, sent the details to SEP and received their acceptance. Thus, a panel of more than forty engineers from different disciplines was finalized for deputation to SEP.

These engineers were permitted to take their spouses along on their deputation to France. The travel expenditure of the deputed engineers and their spouses was met by ISRO, which continued to pay their salaries in India to authorized nominees. Their service conditions in ISRO remained unaltered. In addition, every engineer was paid a subsistence allowance of 2,400 french Franc per month during the period of deputation. The SEP management arranged single-bedroom houses in a residential complex at Vernon, near the work centre. The period of deputation varied from one to four years and was spread over five years. The maximum number of ISRO personnel at Vernon at any point of time was forty.

The work hours in SEP were different each day; the guiding factor was the work hours per week specified by the government of France from time to time. On Mondays, the French started late and on Fridays, they closed early in accordance with their habit of enjoying long weekends. Our members were attached to various teams in SEP and reported to the French engineers, who were quite cordial. The ISRO personnel had been offered language training before their departure to France. The French lessons were taught by an American lady who was married to a Frenchman, and our staff acquired working knowledge of French.

Along with their French counterparts, our engineers were involved in the design and analysis of the components of the liquid propulsion systems, follow-up of fabrication and testing of subsystems and assemblies. The ISRO employees who were deputed under the twenty-five-man-year programme were directly working in the subsystems of the Viking engine, while those under the seventy-five-man-year programme were placed in other areas of the Ariane launch vehicle, specifically working on the stage and ground systems for Ariane's Stage 2. A few ISRO personnel were placed in the area of cryogenic engine and stage developments for Stage 3 of the Ariane vehicle. The deployment helped ISRO to acquire substantial information on the second stage of Ariane and gave the engineers good exposure to its cryogenic third stage.

ISRO personnel taking their spouses to France promoted interaction among the Indians at SEP. The working days were from Monday to Friday, and so the families could spend the weekends sightseeing. Most of the engineers and their families visited other European countries. Support from the embassy was arranged, so they could use the benefits available to all Indian officials on deputation in France. Some ISRO personnel pooled in money and even bought shared second-hand cars, to enjoy the resulting mobility. Whenever senior officials from ISRO visited France, they met our teams and assured them of their support. There was also constructive contact with French families and joint picnics were organized.

Over a period of fourteen years, I executed the liquid propulsion and associated systems development and reported to the director of VSSC. As the person responsible for the control systems of ISRO launch vehicles and satellites in the Auxiliary Propulsion System Unit (APSU) – an autonomous unit of ISRO – I also reported to the chairman of the organization. I simultaneously executed various roles in my capacity as group director, project director and programme director, with multifarious authority and management systems. At times, I had difficulties in my management system. While I had complied with all the orders of the ISRO management, I was not clear about the organization's management strategy.

It would have been pertinent to integrate the two activities under my responsibility – developments in liquid propulsion for launch vehicles, and auxiliary propulsion in control systems for launch vehicles and satellites. Such an integration would be efficient and effective for the development in ISRO. To do so, ISRO had to establish an R&D centre exclusively for liquid propulsion system development with me as the obvious choice to head it. However, as it is in most organizations, office politics played a role here in not establishing such an integrated centre though it affected the developments in propulsion system technology adversely. When the activities of ISRO projects under my responsibility expanded considerably, organizational change became inevitable and, based on my good performance, Liquid Propulsion System Unit (LPSU) was established in 1985 with me as director. It was renamed in 1988 as Liquid Propulsion System Centre (LPSC).

The Vikas–Vernon Project Office

In 1974, I established the Vikas–Vernon project office at Vernon, near the SEP work centre in France, and S. Nambinarayanan, then head of the liquid propulsion division of my PSN group was deputed to France as the project manager. Unni Paramangalath, the administrative officer from my group, was deputed to SEP to maintain and operate the project office, assist the project manager in all administrative activities and provide logistical support to the ISRO personnel there. The project manager interfaced with SEP officials and me on all technical and managerial matters related to the deputation of ISRO engineers, the supply of pressure transducers and the Viking technology acquisition as per the ISRO–SEP agreement.

I decided that the technology acquisition from SEP should be mainly through the ISRO engineers deputed under the 100-man-years programme. Such a step would not require additional expenditure towards technology acquisition. However, more personnel could be deputed to SEP, if necessary. I evolved a technology acquisition methodology and plan, in consultation with everyone concerned, and implemented it. Accordingly, the ISRO engineers at SEP were assigned specific tasks in the technology acquisition scheme, which was coordinated by S. Nambinarayanan under my overall guidance. These assignments were in addition to the tasks allotted by SEP and ISRO under the agreement.

At the same time, to support and implement the technology acquisition programme, I established a backup team in the Liquid Propulsion Project (LPP) as its director at the VSSC. I nominated T.P. Rangamani as the leader of the backup team. In addition, I periodically monitored and reviewed the status of, and progress in, the implementation of the agreement. I normally visited SEP, Vernon, once in six months to check on the technology acquisition and transducer supplies, and to resolve technical and managerial issues, if any. I am happy to state that the ISRO–SEP agreement was implemented successfully in a systematic manner. On completing their respective deputations, the ISRO engineers on the Vikas–Vernon team returned in 1978 and from then onwards joined the backup team at LPP.

Pressure Transducer Supply to SEP

As per the ISRO–SEP agreement, the Vikas Project had to supply around 7,000 pressure transducers as specified by SEP. As the first step towards this goal, a pressure transducer unit (PTU) was established as an autonomous unit of ISRO on the National Aeronautical Laboratory (NAL) campus in Bengaluru in 1974. The PTU was a unique establishment, different from other ISRO units as it had been given the responsibility of producing space-qualified pressure transducers to be exported to France. I shall present a detailed summation of the work there, the import substitution and indigenous developments, and their contribution towards the development of the ISRO's propulsion system in the following chapter.

As the project director of Vikas, I was responsible for the acquisition of the pressure transducer technology from SEP, the production of transducers in PTU and their supply to SEP. In consultation with everyone concerned, I identified fifteen technologies to be acquired from SEP:

1) Precision machining/fabrication
2) Vacuum brazing
3) Gold/copper electroplating
4) Electron beam welding
5) Glass to metal sealing
6) Precision strain gauge bonding, curing and micro-soldering
7) Precision soldering
8) Temperature cycling and compensation
9) Assembly and integration
10) Calibration and proof cycling
11) Vibration and shock testing
12) Acceptance and qualification testing along with insulation resistance and continuity checks
13) Repeatability, stability, sensitivity, non-linearity and hysteresis validation
14) Quality assurance and documentation, and
15) Packaging, export-quality packing, storing and forwarding to SEP with documents.

During the second week of August 1974, I deputed a team of five engineers from PTU to SEP, Bordeaux, France, to acquire technology on three types of space-qualified pressure transducers – 21NA, 21UA and 21UR*. The engineers chosen

* 21NA, 21UA, 21UR are designations depending upon the measurements being absolute or relative.

were N. Gunaseharan (group leader), N. Padmanaban (assembly and calibration in-charge), B.S. Natarajan (testing in-charge), K. Srinivasamurthy (special process in-charge) and Dr M.M. Nayak (assembly and calibration in-charge). This team was accommodated in a laboratory at SEP and provided with stereo microscopes, power supply, soldering station and all the tools needed for the assembly and calibration of transducers for training and production.

Subsequently, more members joined this group to continue the fabrication, testing, electroplating, vacuum brazing, electron beam welding, and inspection and quality control. They were R. Manickavel, M. Balasubramanian, Arunachalam Pillai, K.K. Francis, H.G. Bhaskar, A.S. Nagaraju, Antony Paul Koloa, G. Raman and V. Sunderrajan. The technicians from PTU were trained by SEP experts in machining precision components. An important process in the final transducer assembly was EBW, which was new to PTU. The PTU personnel were trained in all the processes, including EBW, at SEP.

Training started in the most critical areas such as assembly, calibration, testing and integration of transducers by SEP experts – Lachaud, Balzer, Caid, Zerome and Marache, aided by Ms Dumartin, the secretarial assistant. The SEP specialists appreciated how the PTU personnel acquired all the technologies and skills within a short time. This advantage resulted not only in the completion of training within eight months but also in the production of around 100 pressure transducers by the PTU team at SEP, which were then supplied to their programme. It further demonstrated the successful acquisition of pressure transducer technology by PTU.

In August 1975, PTU obtained a new building on the NAL campus in Bengaluru with all the facilities for the production of pressure transducers for SEP. T.N. Seshan provided all the administrative support needed to this priority project of ISRO. V. Srikantan successfully established PTU under my technical and managerial guidance. After returning to India, the PTU team produced, during the first quarter, seventy 21UR transducers and exported all of them before December 1975 – within a year of laying the foundation for the new PTU building. Every year thereafter, PTU could deliver about 800 to 900 transducers of various pressure ranges to SEP without any penalty. This achievement was possible due to the quick responses and support from the top management of ISRO and the committed and sustained contributions of my colleagues.

I wish to express my appreciation for my team members and sincerely thank them. V. Srikanthan shouldered the overall responsibility for the pressure transducer production programme initially and established all the facilities at PTU. N. Gunaseharan, was in charge of all the manufacturing activities in Bengaluru and did so perfectly. He subsequently took over the responsibility of the programme from V. Srikanthan when the latter left ISRO. N. Padmanabhan oversaw the transducer assembly, testing and calibration activities, and R. Manickavel, looked after the inspection and quality assurance activities for all developments in Bengaluru.

During the formative stages of PTU, I travelled to Bengaluru once a week, and monitored and reviewed all activities, including production and delivery status. At times, I participated in testing and calibration, in addition to doing the regular administration work and corresponding with our SEP counterparts. Since PTU did not have a vibration table, on many occasions I carried the assembled pressure transducers to the VSSC, made arrangement for conducting the vibration tests, and sent them back to PTU in order to meet the delivery schedule.

System and Procedure for Pressure Transducer Supply

Jointly with SEP, I established the following system and procedure for the supply and acceptance of transducers. The supply of pressure transducers to SEP was to be carried out at every quarter. To organize their manufacture and supply on time, SEP would provide details to ISRO two quarters in advance – including the numbers, the type, the pressure range and the specifications of transducers to be delivered every quarter. Accordingly, ISRO would then manufacture and ship these transducers to SEP, along with the specified documents and calibration reports stipulated in the agreement. The SEP would review these documents and also inspect the transducers at random in their laboratory before they accepted the shipments. In case SEP could not accept any of the transducers due to non-conformity, we could discuss the details in joint meetings, sort out the issues and take remedial action.

Apart from maintaining regular communication, representatives of the two organizations were to meet at least once in six months to review the status of supply and resolve any problems. I recall that we introduced some modifications in the manufacturing processes and inspection methodology during the initial stages. I am pleased to say that PTU completed the supply of more than 7,000 pressure transducers of different types and pressure ranges on time as per the ISRO–SEP agreement without incurring any penalties.

This feat was made possible by the dedicated contributions of my colleagues.

Liquid Propulsion Projects

Consequent upon the ISRO–SEP agreement, with its focus on the development of high-thrust, high-performance liquid engines for the ISRO launch vehicle programmes, an integrated institutional set-up was necessary to implement the technology development in a project mode. S. Nambinarayanan suggested to the chairman of ISRO that he be made the chief of the project to implement the Vikas engine realization in India. All members of the backup team, who were working with me for hardware realization and establishing test facilities, were to be placed under him. Considering the organizational system that I had evolved – which assigned tasks to senior engineers with responsibility and accountability, and had proved successful in the progress already made – the senior engineers working with me requested the administration to permit them to continue under me with the existing management system. I had a joint meeting with S. Nambinarayanan, K. Kasi Viswanathan and R. Karunanidhi, in which we frankly discussed everyone's views. I later communicated the outcome of the deliberations to the chairman of ISRO.

After examining several factors, Dr Dhawan established the liquid propulsion projects (LPPs) in 1980. I was made the director and there were three senior colleagues, all from the propulsion group in the VSSC, with major responsibilities as outlined below:

1. S. Nambinarayanan was responsible for all activities related to Vikas engine and stage developments.

2. K. Kasi Viswanathan was accountable for all activities related to Vikas engine hardware realization.
3. R. Karunanidhi was in charge of all activities related to the assembly, integration and testing of Vikas engine and stages. In addition, he was responsible for establishing the infrastructure at the Liquid Propulsion Test Facilities (LPTF) in Mahendragiri, a new site.

Subsequently, S. Nambinarayanan met me alone and expressed his regret for rebelling against me and writing to the chairman, and for claiming the top executive position without consulting his senior colleagues. I told him that I appreciated his ambitious nature and explained the facts and circumstances in the interest of our institution, which he had failed to understand fully. At our meeting, he pledged his support and cooperation as he had done in the past. I ignored the unpleasant events and chose to have him as my deputy.

I used to have periodic joint meetings with S. Nambinarayanan, K. Kasi Viswanathan and R. Karunanidhi to discuss plans and follow-up actions. At times, I met them individually, whenever the need arose. We worked together as a team to carry out these activities.

Vikas Engine and PSLV Stage Development: Technology Acquisition

The development of the Vikas engine and stage for PSLV was based on the technology acquired from SEP under the ISRO–SEP agreement. In consultation with everyone concerned, I had finalized the technology acquisition programme, which was monitored and executed on a day-to-day basis under the leadership of S. Nambinarayanan, the project manager of Vikas–Vernon. ISRO engineers interacted with specialists from SEP and other European industries. The technology transfer agreement was for the Viking engine but ISRO engineers were associated with the Viking stage developments at SEP. The latter included inter-stage thrust frame, propellant tanks, pressurization system, fill-and-drain system and gimbal control system, among others. The exposure and knowledge gained from SEP on the stage elements were valuable for the liquid stage development at ISRO. Our engineers held discussions with SEP specialists, who provided clarifications and additional information to gain the 'know why' in addition to the know-how.

I appreciate the leadership of S. Nambinarayanan and the commitment of ISRO engineers to the different subsystems of the Vikas stage assigned to them under the technology acquisition scheme. This team included P. Mohanaprasad, V. Gnanagandhi and K. Manikantan in engines; M.K.G. Nair, E.S. Joseph and Baby T. Mathew in components and modules; K. Lakshiminarayanan, M.K. Narendranath and P.P. Kailasanathan in test and evaluation; and G. Krishnamurthy, P.S. Krishnamurthy and D. Anandan in fabrication, assembly and integration; along with others in project management, quality assurance and so on.

Vikas Engine Hardware Realization

The next step in the Vikas project was the realization of the Vikas engine hardware in India through our industries under K. Kasi Viswanathan, project manager of fabrication, LPP. I proposed that we manufacture the major components of the Vikas engine initially using dummy materials based on the documents and drawings received from SEP, and generated by the Vikas–Vernon team and the backup team at LPP. Towards this objective, LPP established fabrication contracts with a minimum of two industries for every major component, such as injectors, thrust chambers, engine valves and component modules. K. Kasi Viswanathan invited potential industries, including MTAR Technologies, Hyderabad, Godrej & Boyce Pvt Ltd., Mumbai, Cooper Engineering, Pune and Walchandnagar Industries, Mumbai, and explained the details of the components to be manufactured and the LPP requirements.

In exceptional cases, when facilities did not exist in the industries for the fabrication of special components such as injectors, LPP established in-house units. Having finalized a plan to realize the hardware for the Viking engine model, offers for fabrication were invited from industries that had the requisite potential. Negotiations were held and two firms were selected to fabricate the same component, so as to provide options. Fabrication orders were placed, drawings and dummy materials were supplied, and the hardware realization progress was closely monitored. After the successful completion of fabrication, the better of the two samples was selected after an assessment.

In this process, I recall that both MTAR Technologies and Cooper Engineering manufactured the engine valves and, finally, MTAR Technologies was chosen to proceed further. MTAR Technologies and Godrej & Boyce were also selected to manufacture the thrust chamber. The main injector was initially fabricated in-house but MTAR Technologies was picked for the purpose. It was also selected for turbopump components and gas generator. All the components of the Vikas engine – with the same dimensions as the SEP engine – were manufactured with indigenous materials and assembled successfully in 1978.

Vikas Engine Realization

After we had established the facilities and demonstrated the fabrication technologies in industries for the manufacture of the Vikas components, I approached Dr Dhawan for approval to import three sets of all the materials and ready-to-use, certified items, such as bearings, seals and gaskets, from the same party that supplied to SEP when they were making Viking engines for the Ariane Project. It would ensure quality and eliminate testing and acceptance process for use at ISRO. The chairman approved this request and made the funds available as a special case.

These purchases were made through SEP and coordinated by the project manager of Vikas there. Thus, in 1982, three sets of materials and bought-out items in accordance with the SEP specifications were available in India. The project manager of LPP fabrication and his team initiated the realization of the first Vikas engine with imported materials, using in-house facilities and with involvement of external industries that supplied components.

In 1984, I was extremely happy to see the first Vikas engine with imported materials, manufactured totally in India and assembled successfully. This achievement was possible due to the unrelenting efforts of my colleagues K. Kasi Viwanathan, R. Karunanidhi, S. Nambinarayanan, A. Ramasubramanian, D. Anandan, H.R.S. Mani, T. Jeyakumar, V. Gnanagandhi, G. Krishnamurthy, P.S. Krishnamurthy, K. Lakshiminarayanan, A. Chandran, K. Manikantan and their team.

Based on the contribution of the industries, LPP gave the contract for the total Vikas engine as a unit to three different organizations – MTAR Technologies of Hyderabad, Godrej

6. First bi-propellant rocket with 600 kgf thrust engine at the launcher for flight test at Shriharikotta launch pad

7. First bi-propellant single-stage liquid rocket with 3000 kgf thrust bi-propellant engine on launcher (left) and in flight (right) at Shriharikotta launch station

8. Schematic diagram of the cross section of Viking engine with thrust chamber, turbo-pump, gas generator, valves and flow control system

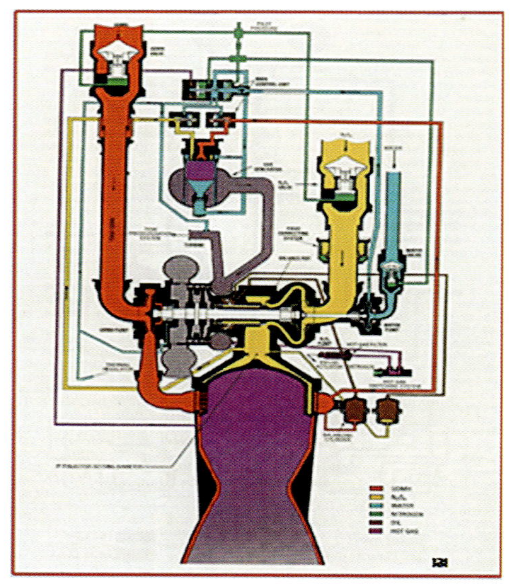

9. Vikas liquid engine with improved performance, designed, developed and qualified for ISRO launch vehicle programme by the Liquid Propulsion Systems Centre (LPSC), in association with Indian industries, based on the Viking engine technology from Europe

10. Second stage of PSLV integrated with Vikas liquid engine for launch in PSLV- C10 flight.

& Boyce of Mumbai and Kerala Hitech Industries Limited (KELTEC). They delivered their first Viking engines in January 1992, in December 1993 and in 1996, respectively. These three agencies continued to supply Vikas engines (Photo Nos 9 and 10) regularly for all future ventures of ISRO.

Vikas Engine Test at SEP

In 1982, LPTF was established as a unit of LPP at Mahendragiri, with R. Karunanidhi as project manager who reported to me in my capacity as the director of LPP. The principal test stand, one of the major test facilities to test the Vikas engine and stage, was not ready by 1984. I suggested that the first India-made Vikas engine be tested at SEP facilities. Dr Dhawan approved of the idea and authorized a visit to SEP for T.N. Seshan, V. Gnanagandhi of LPP, A. Chandran of the ISRO headquarters and me. We were supposed to discuss with the SEP authorities and finalize a test plan to conduct a full-duration test of the Vikas engine for 120 seconds (nominal). Accordingly, along with the project manager of Vikas–Vernon, we held a discussion with the SEP officials who informed us that they were busy with the committed test programme for the Ariane Project. Further, SEP expressed apprehension that in the unlikely event of a major failure of the Indian Viking engine, the test stand could get damaged and the SEP commitment to the Ariane programme would be adversely affected.

In response, I pointed out that the Vikas engine was based on the design and documents of SEP. I also suggested that SEP could depute experts to ISRO to verify all the processes and inspection documents of this engine, and hold discussions with the engineers associated with the hardware realization activities after their training at SEP. If necessary, they could carry out additional inspections until they were convinced of the quality of the hardware. In addition, the total hardware could be shipped to SEP and assembled in the presence of SEP experts. If the latter were satisfied, a short test of ten seconds, followed by a single full-duration test of 120 seconds could be carried out by SEP personnel. All the support services such as propellant servicing, gas filling and instrumentation should be provided by SEP.

The SEP proposed that the test be carried out in two phases. The first would be testing and evaluating the performance of the gas generator. The second phase was integrating the gas generator with the Vikas engine and conducting a short-duration test, followed by a full one. Both ISRO and SEP agreed to the proposal.

Accordingly, SEP deputed a team of its experts to LPP in 1985. V. Gnanagandhi of LPP was to organize the visit of the SEP team. The SEP team reviewed all the processes and inspection documents, held discussions with the ISRO fabrication team and examined the critical components. They also visited the facilities of the industrial partners – Godrej & Boyce in Mumbai and MTAR Technologies in Hyderabad. The SEP team was pleased with the quality of the fabrication work in these centres. In the end, all of us in the Vikas backup team were extremely happy to hear that SEP was satisfied with its evaluation and that ISRO could proceed with the proposed test of the Indian Viking engine.

An agreement was executed between SEP and LPP. A team of about fifteen personnel, under the leadership of S. Nambinarayanan, was identified to implement the agreement. The first visit of the ISRO team to SEP for the first phase of testing the gas generator took place during the second half of 1985. However, the gas generator test facility at SEP was not in operation since the testing of the SEP gas generator for the Ariane programme was completed quite some time ago. The first test of the ISRO gas generator was a failure because the filters in the line got contaminated. The SEP and ISRO team jointly analysed the problem and concluded that the gas generator facility had contamination that blocked the filters and that the failure was due to the contamination and not due to any hardware issue. The fluid circuits in the facility were thoroughly cleaned, the filters were replaced, the ISRO gas generator was reassembled and the gas generator successfully tested. The performance was as per design. The ISRO team members returned with smiling faces.

Under the leadership of S. Nambinarayanan, the team went to SEP again in December 1985 for the second phase of the test programme. As planned, short-duration and full-duration (150 seconds) tests were conducted at the SEP, Vernon, main test stand on 9 December 1985. The test was flawless; the performance of the Vikas engine was excellent and compared very well with its SEP counterpart. Thus, a major milestone in the Viking engine at ISRO was successfully demonstrated. Dr Dhawan and many senior officials congratulated LPP on the grand success.

Vikas Engine Supplies to ISRO

Along with the supervision of the hardware manufacturing programme of the Vikas engine in Indian industries, I closely monitored the progress in establishing the assembly and test facilities in a phased manner at Mahendragiri. In 1987, when the minimal essential facilities were commissioned, the LPP carried out the first Vikas engine test at the LPSC in

Mahendragiri, successfully. This test was a major milestone in the history of the development of the liquid propulsion system in ISRO under my leadership.

I was pleasantly surprised when ISRO called me the 'Father of ISRO Liquid Propulsion', in recognition of my contribution to the programmes. Let me once again place on record the sincere and unrelenting efforts of my colleagues – S. Nambinarayanan, K. Kasi Viswanathan, R. Karunanidhi, V. Gnanagandhi, K. Sivaramakrishnan Nair and their teams over the years. I appreciate their inputs and thank them all for the excellent work that culminated in the spectacular success.

Around the same time, I initiated programmes for material indigenization and performance enhancement. With the pedigree of the efficient performance of the Vikas engine in ISRO's missions, the uprating of the engine was taken up with minimum modifications to improve the payload capability of the launch vehicles. I am happy to note that these high-thrust Vikas engines have been fruitfully inducted into the ISRO programme. Moreover, the production of Vikas engines has been effectively implemented through a consortium of Indian industries. I am thrilled to see that today, the LPSC supplies Vikas engines with improved performance for use in PSLV, GSLV and GSLV-Mk-III.

6

Pressure Transducer Unit: Indigenization, New Developments and Technology Transfer

As DISCUSSED IN THE last chapter, in 1974, a Pressure Transducer Unit (PTU) was established as part of the Vikas project, as an autonomous unit of ISRO on the NAL campus in Bengaluru. In addition to fulfilling the ISRO–SEP agreement, the PTU also established special fabrication facilities and manufactured thrusters and propellant tanks for the ISRO satellite programmes. In addition to producing pressure transducers as per SEP requirements, I established a team under the leadership of Dr M.M. Nayak for indigenization and development of various transducers.

PTU supplied these pressure transducers to the specifications of the ISRO launch vehicle and satellite projects and to other national laboratories and industries. NIO, Goa, NAL, Bengaluru, BHEL, Tiruchirappalli, BHEL,

Vikas Nagar in Hyderabad, ISM, Dhanbad, ADE, Bengaluru, DRDO, Hyderabad, IISc, Bengaluru (especially its civil engineering, aerospace and instrumentation departments) and SSPL (DRDO) in New Delhi were prominent among them. Dr Nayak's team developed other types of transducers too – differential pressure transducers (Photo No. 11), temperature transducers (Photo No. 12), integral diaphragm pressure transducers for satellites and launch vehicles, Microelectromechanical System (MEMS), 1 bar pressure sensors, and MEMS differential pressure transducers among others. The team also came up with sensors for automatic weather stations, platinum-wire-wound temperature sensors, thermocouple temperature sensors, ultrasonic level sensors, capacitance level sensors and splashdown sensors, to name a few.

Today, PTU has successfully transferred the pressure transducer technologies to the following ten institutions:

1. Systronics, Bengaluru
2. Macurex Sensors Pvt. Ltd, Bengaluru
3. Keltron, Thiruvananthapuram
4. Keltec, now known as BrahMos Aerospace Trivandrum Ltd (BATL), Thiruvananthapuram
5. Shreyas Engineers, Bengaluru
6. Lakshmi Technology and Engineering Industries (LTE), Coimbatore
7. Fabrica Industries, Mumbai
8. Semiconductor Laboratory (SCL), Chandigarh
9. Central Tool Room and Training Centre (CTRTC), Bhubaneswar, and
10. DRDO, Hyderabad.

These organizations are PTU's major industry partners that supply transducers in bulk at short notice. Some of them have established end-to-end capabilities in the business.

PTU Fabrication Facilities

The fabrication facilities established at PTU in 1975 were meant to exclusively produce and supply pressure transducers to SEP. The unit had general purpose lathes and precision tool room lathes (Crouzet TC-125), milling machines, special process facilities such as the 1 cu. ft 30kV/3kW EBW machine, the Balzers Vacuum Brazing and Muffle furnaces, and a gold-plating facility among others. During the period 1975–85, the unit was given additional fabrication responsibilities producing spacecraft propulsion systems components, which included 1N monopropellant thrusters and 30-litre-capacity propellant tanks with bladders (Photo No. 13) for the IRS class of satellites.

Moreover, the machining of parts for INSAT-class satellites, such as microwave filter cavities and wave guides with Invar-36 thin-walled sections of 0.8 mm, was taken up. The fabrication facilities were augmented with high-precision Schaubline 150, 125, 102 lathes, Moore Jig boring machine, Struder cylindrical grinding machine, Ewag tool and cutter grinding machine, Wasino LG 71 CNC lathe, Jones and Lamson CNC lathe, Sodik CNC EDM from Japan, twin-chamber EBW machine from TORVAC, UK, VFS Vacuum Furnace and an anodization facility.

Further, as fabrication demands increased, the PTU was given the additional responsibility of fabricating components for launch vehicle propulsion systems, which included PS4 titanium alloy propellant tanks, titanium gas bottles for

PSLVs, PS4 nozzles made of columbium alloy divergent (Photo No. 14) and PS1 RCS propellant tanks. The EBW technology was developed for welding thin-walled titanium alloy surface tension type spacecraft propellant tanks, and in welding dissimilar columbium alloy (new name niobium) C-103 and titanium alloy Ti-6Al-4V. The fabrication facilities were further augmented with Deckel CNC milling, Monforts and Weiler CNC lathes, high voltage EBW facility with a 27 cu.m vacuum chamber and 150kV/15kW EB gun from Cambridge Vacuum Engineering (CVE), UK, electroplating facilities, etc.

To meet the increased demand for thrusters and flow control valves for 1N, 11N propulsion systems, propellant management devices (PMDs), propellant acquisition systems (PAS) for PS4-PSLV, 21 NA pressure transducers and so on, the fabrication of components was outsourced to external agencies. A heavy-duty Innse CNC lathe was placed at Bharat Heavy Electricals Ltd (BHEL) in Ranipet, Tamil Nadu, to support the fabrication of specialized propellant tank parts. Apart from utilizing the industries to supply the components required, ISRO entered into end-to-end production contracts with those enterprises that had demonstrated the capability to supply 21 NA pressure transducers, differential pressure transducers and about fourteen types of spacecraft propulsion system components. In addition, the government-owned, company-operated (GOCO) model was introduced for the supply of spacecraft propulsion system components. Accordingly, an integrated titanium alloy tank production facility was planned for the end-to-end production of titanium alloy propellant tanks for spacecrafts and the upper stages of launch vehicles.

11. Differential pressure transducer developed by the Pressure Transducer Unit (PTU)

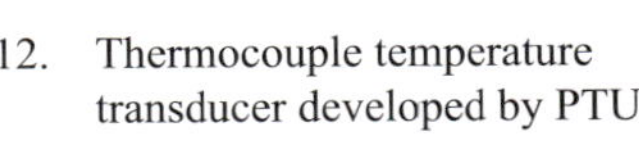

12. Thermocouple temperature transducer developed by PTU

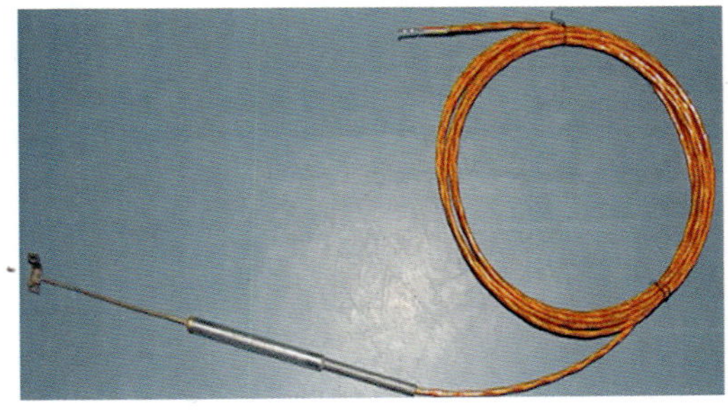

13. 30-litre capacity, thin-walled propellant tank for Indian Remote Sensing Satellite with positive expulsion device

14. Director, LPSC and GM, PTU, inspecting the successful electron beam welding process for welding large expansion divergent nozzle made of a Columbium alloy, for the regeneratively cooled liquid engine for PSLV fourth stage

15. Professor Satish Dhawan, chairman, ISRO, T.N. Seshan, joint secretary, DOS, Dr A.E. Muthunayagam, founder-director, LPSC, S. Nambinarayanan, Vikas-E&S, N. Gunaseharan, GM, PTU along with Rane Morin, commercial director of the French aerospace company SEP and his team after final review of transducer supply agreement between ISRO and SEP

16. Members of the PTU team at Bangalore

It is heartening to observe that the PTU has met all the SEP commitments on time, has made significant contributions in transducer technology development and has fulfilled the needs of the ISRO launch vehicles and satellite programmes. Further, it has expanded the fabrication facilities and manufactured sophisticated high-technology components for ISRO programmes. The transfer of technology to Indian industries so that the latter can contribute to our space endeavours is praiseworthy (Photo Nos 15 and 16).

7
Solid Propulsion Systems: SLV-3 and Beyond

I WAS ACTIVELY ENGAGED in, and responsible for, the development of solid propellant motors for all the ISRO programmes during my initial days there until I took over the responsibility to develop liquid propulsion systems for both launch vehicles and spacecrafts. Subsequently, the solid propellant motor developments were carried out jointly by the solid propulsion division of my propulsion group (with Rajaram Nagappa as the divisional head), the System Development Agencies (SDAs) of launch vehicle projects (SLV, ASLV, PSLV and GSLV), and propellant manufacturing agencies – Solid Propellant Space Booster Plant (SPROB) and Rocket Propellant Plant (RPP). I was associated with these ventures as a guide, supervisor and reviewer. A brief summary of solid propellant motors developed for all the programmes of ISRO during my tenure and subsequently is presented below.

Solid Motors in SLV and ASLV

Solid motor technology in India is totally home-grown, right from the era of the Rohini sounding rockets of the 1960s. INCOSPAR, the predecessor of ISRO, started launching indigenously realized sounding rockets in 1965 and the expertise gained was of immense value. The experience obtained in the design, processing, testing and integration of the Rohini (RH) series instilled the confidence necessary for the design and development of stage motors for SLV and ASLV, with all-solid propellant motor configuration. The SLV had four stages, with a boosting thrust of 45 tonnes provided by a 1 m diameter solid motor (S1) having a high-strength metallic casing in segmented configuration. The second and third stages (S2 and S3) had an 800 mm diameter monolithic motor case. The terminal S4 motor was initially designed for 560 mm diameter, which was later increased to 647 mm, in keeping with the requirements of CNES (Centre Nationale d'Etudes Spatiales, the French Space Agency).

The first and second stages (S1 and S2) were loaded with a PBAN (Polybutadiene Acrylonitrile)-based propellant named RCN-01, and S3 and S4 were loaded with a CTPB (Carboxyl-Terminated Polybutadiene)-based high-energy fuel HEF-20 propellant. For the design of stage motors for the satellite launch vehicle, the limited literature available on the SCOUT (Solid-Controlled Orbital Utility Test system, NASA) launch vehicle was taken as the reference. Taking into consideration the propellant processing capacity of RPP, which was commissioned in 1972, the S1 motor was designed as a segmented motor with a 1 m diameter.

The propellant manufacturing facilities at the Solid Propellant Space Booster Plant (SPROB) in Sriharikota were then in the design phase. I recall that developmental issues, such as wide dispersion in properties, curing cycle and cracks in the propellant grain at the port internal diameter, were studied in depth and corrective measures successfully implemented during subsequent casting trials at SPROB. An attempt was also made to go for a monolithic propellant grain (without segment joint) for the S1 motor.

After establishing the casting facilities at SPROB, the team processed a monolithic motor by casting 9 tonnes of propellant, instead of 3 tonnes each in three segments. The motor was static-tested in 1977. Unfortunately, the test was not successful and so the design was not pursued further. Instead, there was a proposal to develop an S4 and to supply it as the Apogee Boost Motor (ABM) for the APPLE mission. However, the European Space Agency (ESA) demanded the adoption of the batch processing concept for ABM. The SPROB team successfully developed and demonstrated batch processing, which was adopted later for ASLV motors.

The nozzle employed for all the motors was simple, with a configuration of rigid Convergent-Divergent (CD), externally mounted to the motor case at the rear end. The Propulsion Engineering Division (PSN) under my responsibility took the lead in the design and development of nozzles. The material used was graphite for the throat region, and carbon/phenolic and silica/phenolic for the convergent and divergent regions, respectively, with high strength 15CDV6 as the structural backup. In those days, the flex nozzle technology was not thought of for controlling thrust vector as the mission demand was limited to within ±2° thrust vectoring. Hence, the secondary injection thrust vector control system was used, along with the Fin-Tip Control System.

To increase the payload capability from 40 kg in LEO (Low Earth Orbit) to 150 kg, ASLV was configured with all solid stages of SLVs by adding one more stage in the boost phase. A simpler configuration, by strapping two identical motors, which were similar to S1 motor, with the core S1 stage, was selected to avoid a fresh development of new motors that would entail additional effort, facility and time. However, the main configuration change was canting the nozzle of the two strap-on motors by 9° to orient the thrust vectors of the strap-on motors to pass near the centre of gravity of the vehicle.

I recall the dedicated developmental efforts of my propulsion division, the Solid Motors Project (SMP), all SDAs and processing facilities as they all strived to develop quite a few new technologies and improvements to meet the mission needs. I congratulate the leaders and members of the teams for their untiring work.

The experience gained by the solid motors team from the design, analysis, processing and integrating to testing was of immense value. I look back on those challenging tasks, the performance of which encountered many setbacks in the initial stages of solid motor static testing. The obstacles were taken as lessons for and stepping stones to success. Though we had many failures in static tests during the development, no solid motor of ISRO was unsuccessful in the launch vehicle mission, except a single, rare incident of non-ignition in one of the ASLV missions.

Solid Motors in PSLV

During the Polar Satellite Launch Vehicle (PSLV) configuration study phase in 1978–79, the mission objective given to the launch vehicle team was to launch an IRS class satellite that weighed 900 kg into the 900 km polar orbit. The finalized configuration was with a 2.8 m diameter core vehicle having six solid motor strap-ons of proven 1 m class. The first and third stages were solid motors, while the second and fourth stages were with earth-storable liquid engines. The two solid stages were based entirely on new designs, with a quantum jump in size and thrust capability, which demanded the state-of-the-art technology, new material, processes and facilities, both in-house and at external work centres.

The Design and Development of the S125 Motor

The first stage motor PS1, named S125 booster, provides the main lift-off thrust to PSLV, along with either two or four strap-ons out of six, and lifts the vehicle to an altitude of 76 km, thereby imparting a velocity (ΔV) of around 2 km/sec. The S125 was designed with 129 tonnes of HTPB (Hydroxyl-Terminated Polybutadiene)-based propellant loaded in five segments of 2.8 m diameter and 3.4 m length each. The motor case was designed with high-strength M250 grade maraging steel, which was developed in collaboration with Mishra Dhatu Nigam Ltd (MIDHANI).

The S125 motor case is made of five segments with two dome-end segments – namely, Head End Segment (HES) and Nozzle End Segment (NES), and three Mid Segments (MS). The segments are with shells rolled and welded construction with forged end rings welded to shells. The tang-and-groove joint is configured with capture features to prevent the rotation of the joint under operating pressure, and pressure sealing is ensured by one face and one shaft O-ring of Viton® rubber. I recall that the Space Shuttle Challenger disaster of

1986 had alerted our designers to improve the joint with the introduction of capture features. All the segments are individually proof pressure tested to 66 ksc* (1.1 times the Maximum Expected Operating Pressure – MEOP – of the motor).

Propellant grain at the HES is designed with ten deep, lobed stars to enable high take-off thrust and also to minimize the dynamic pressure, whereas in all the other segments, grain geometry is cylindrical with an interchangeability option within MS. This grain configuration provides typical M-type thrust-time curve with high take-off thrust. The grain end faces in MS and HES aft ends and NES fore end are inhibited with IR-1 resin to prevent end face burning (Photo No. 17). Segment joints are provided with predetermined compression at the IR (inhibition resin) faces coated with a potting compound called PC-10. Thermal protection for the motor case is provided by a proven nitrile-based rubber formulation called ROCASIN (Rocket Case Insulation) and cured in autoclave at 120°C with 5 ksc pressure. ROCASIN at the segment interfaces and HES and NES openings, which are provided with tang and groove, are filled with PC-10 during the segment integration.

The igniter used is two-stage pyrogen with the primary igniter ignited by a secondary igniter, which is, in turn, ignited by a booster charge triggered through an ignition train from Remote-Mounted Safe Arm (RMSA), Explosive Transfer Assembly (ETA), manifolds and Through Bulkhead Initiators (TBIs), which are mounted at the secondary igniter end dome. The nozzle is fixed with external convergent-divergent configuration employing carbon-phenolic and

---

* ksc=kg/sq.cm, a unit of pressure.

silica-phenolic ablative liners for erosion and thermal protection. The nozzle has the provision to mount twenty-four Secondary Injection Thrust Vector Control (SITVC) valves at divergent fore end to inject strontium perchlorate through blowdown mode based on the command from the control system to enable thrust vector control in pitch and yaw directions.

The following are some of the new technologies developed and facilities established at the VSSC, the SDSC and in external work centres for the realization and qualification of the S125 motor:

- Establishment of the facilities at work centres such as rolling and welding stations, heat treatment, machining and proof pressure testing facilities.
- Development of non-destructive testing (NDT) techniques and evaluation of fracture toughness.
- Development of filler wires for M250 welding.
- Realization of tooling and fixtures, including the jigs, to enable the interchange of segments realized at different work centres.
- Development of M250 segments carried out through a special programme called maraging steel indigenization programme.
- Propellant development, scale-up and mixing in change-can vertical mixer, realization of complicated casting mandrel with deep-star configuration, process hardware and Teflon coating of the mandrels, insulation lining and machining techniques.
- Commissioning of 2 m diameter hydroclave for high pressure curing of ablative liners, special techniques

for the processing and curing of large-diameter nozzle liners and the commissioning of 2 m diameter Velocity Trimming Module (VTM) for liner machining.

- Adoption of carbon/phenolic in place of graphite as the throat material.
- Design, development and qualification of a two-stage pyrogen igniter with redundant RMSA unit.
- Segment assembly in vertical condition and integration techniques with adequate thermal and pressure sealing with face compression between the segments.
- Design, realization and establishment of the horizontal static facility 6C at STEX with thrust wall of 750-tonne capacity, crane capacity of 175 tonnes and a six-component test stand with a provision to measure the side thrust.
- Segment assembly, nozzle and igniter assembly with the necessary thermal barriers and testing with the SITVC duty cycle required.
- Subscale tests at 1 m diameter motor level, viz., S1 dry joint motor and S3 proof motor for ballistic performance evaluation and performance evaluation of the thermal protection system (TPS) and interface elements, especially segment joints, have given valuable inputs and confidence for the realization of two motors for static tests.

An expert team was appointed under my chairmanship to be associated with the entire development programme through reviews at different phases of development from design to the static test of the motor. The other members of the team included Rajaram Nagappa, his colleagues – Dr S.K. Radhamohan, T.T.A. Tharakan, H. Somasundaram,

R. Sundaresan, S. Lakshiminarayanan, R.V. Perumal, Hanumantha Rao, K. Harinath, P.P. Sinha, S.V. Sharma, S. Murugesan and Eswaran V. Gounder – and other young engineers. The deliberations of the committee took over six months and were very transparent in giving inputs whether it was a setback or a success that had made the review process effective.

I recall that the team had contributed substantially to the realization of the S125 motor segments by solving various technological issues that had occurred during the different stages of development. I appreciate their contribution and congratulate them on their achievement. The systematic developmental effort, timely reviews and efficient decision making gave excellent results, with a successful performance in the first S125 static test S125-ST01 carried out on 21 October 1989. Rajaram Nagappa, project director of SMP, was very particular that I should witness the first test of PS1. He made all the arrangements for me to reach the block house at STEX, where the test operations were conducted. I still recall with excitement my adventurous arrival on the day of the test, just prior to the ignition of the motor (Photo Nos 18,19 and 20).

The successful static test of S125-ST01 was a great achievement as it symbolized the fruition of totally home-grown technology and matured the design development of the solid motor with 129 tonnes of propellant with a quantum jump from a 1 m diameter motor. I recall that the magnitude of the R&D efforts involved was significant in comparison with the 1 m diameter motors for SLV/ASLV, with respect to size (1 m to 2.8 m), the development of the case material with almost twice the strength, propellant loading (3 tonnes in a segment to 25 tonnes), thrust capability (45 tonnes to 360 tonnes) and action time (50 to 100 seconds).

The second static test was carried out successfully on 23 January 1991, with minor improvements based on the post-test observations of ST01, especially in the segment joints and motor/nozzle and motor/igniter interfaces. The main objective of the second test (ST02) was to reconfirm the performance of the motor and its subsystems and qualification of the SITVC system with the duty cycle for the maximum side force of 20 tonnes. The performance of the motor in ST02 was excellent and consistent with that of ST01, thereby validating the identical process parameters followed with adequate quality control and assurance measures, which had given enough confidence to certify the third motor for PSLV-D1. Though PSLV-D1 flight was unsuccessful due to software and control system anomalies, the performance of S125 and other Solid Rocket Motors (SRMs) was excellent and as predicted.

Segment Joint and Nozzle and Igniter Interface with Motor

I would like to bring out some of the lessons learnt from the S125 motor development and the results of the S125-ST01. In solid motors, the interfaces – segment joints, nozzle/motor and igniter/motor interfaces – are potentially vulnerable zones as any leak path will lead to a catastrophe. So utmost care is essential to study all possible failure scenarios while designing the interface; therefore, adequate defence on thermal and structural load must be considered.

As I have mentioned earlier, in addition to two O-ring sealings, the incorporation of the capture features at the right time of the development made the segment joint more reliable by controlling the joint rotation due to operating pressure. During the segment assembly, the inhibition to compression at the grain end face was aimed at 5 mm at the port ID (internal diameter) and 2 mm at the grain OD (outer diameter). To ensure this factor, the gap between the insulation tang and groove was filled with potting compound PC-10 as the thermal barrier. The entrapped air between the segments had resulted in a gas leak path till tang and groove of insulation and soot deposits were noticed. However, the face O-ring and shaft O-ring were intact and free from any traces of gas entry. Based on the post-test observation of ST01, a few improvements were made and implemented in ST02. The design outlined above was frozen and flown in PSLV-D1, D2 and D3 successfully.

S125 to S139 with Additional Propellant and TPS Optimization

Subsequently, to enhance the payload capability of PSLV, the S125 design was revised with additional propellant loading of 9 tonnes and designated as S139. Based on the S125 static test performance, there was scope for optimizing the thermal protection system (TPS) for motor case and nozzle design. The design was optimized and the S139 was successfully qualified in S139-ST01. The S139 is being used from the PSLV-C1 mission onwards. It is a workhorse booster for PSLV and GSLV missions.

Design and Development of the Third Stage of PSLV (PS3)

The third stage of PSLV was designed with several firsts and new state-of-the-art technologies to realize this stage with a higher propellant mass ratio (similar to the Inertial

Upper Stage [IUS] of the US). The motor had 7.3 tonnes of propellant, a 2.0 m diameter motor case wound with high strength (130kgf/mm^2) and lightweight (density: 1.4 gm/cc) Kevlar-49/Epoxy motor case with equal length. Unlike the other motor cases, it was provided with a single skirt at the head end, and hence, there was a hanging stage from the vehicle inter-stage structure. The TPS was provided by rocket case insulation (ROCASIN), with a density of 1.2 gm/cc. Initially, the lightweight TPS with cork-filled Ethylene Propylene Diene Monomer (EPDM) formulation having a density of 0.8 gm/cc was attempted. However, due to some development issues during the scale-up trial, it was not pursued further.

For this stage, the submerged and contoured flex nozzle with an area ratio* of 50 was designed and developed by incorporating state-of-the-art technology. With an area ratio of 50, the Specific Impulse (Isp) of 294 seconds could be achieved. It may be noted that the payload sensitivity of the upper stage is high. In the case of PS3, every 1 second Isp would give an additional payload of 7 kg. Similarly, every 2.5 kg reduction in inert mass and the addition of 14 kg of propellant can give an additional 1 kg payload. Flex seal and flex nozzle design was conceived with the limited literature data available and it was a technological challenge that the development team at SMP, Ablative Systems Division (ABSD) (erstwhile PSLV Nozzle Development Facility [PNDF]) and the analysis team at the structures division (STR) faced. The flex nozzle control system with electromechanical actuator, having the stall load capacity

---

* Area ratio is the ratio of the area of the nozzle at the exit to the area of the nozzle at the throat.

of around 2,000 kgf, was to be developed to overcome the stiffness of the flex nozzle and to achieve ±2° vectoring in actuator plane. Two actuators mounted 90° apart could achieve ±2° in omni axis and ±2.8° in the resultant plane.

The design and development of the PS3 motor had the following tasks:

- The realization and qualification of the Kevlar/Epoxy motor case of 2 m diameter for the pressure (MEOP 60 ksc) and vehicle loads (axial compression, tension and bending moment). The design and processing were a challenge as it had unequal opening at HE (200 mm) and NE (750 mm) to accommodate the igniter and submerged nozzles. Several winding trials had to be carried out to freeze process parameters without causing fibre slippage. The team led by C.R. Satya, P.K. Bagchi, M. Ramakrishanan and Ramachandran Nair played a crucial role in the successful development of such a motor case.
- To avoid the degradation of Kevlar/Epoxy material, the insulation was cured at 100°C instead of 120°C, which was normally followed for the metallic case insulation. The HTPB-based propellant with 86 per cent solid loading and 18 per cent Al (aluminium powder used in the propellant composition formulation) was qualified in S3 proof sub-scale motor. The propellant was cured at 50°C, adopting the pressure cure technique (3 ksc) to minimize the cure strain in the composite case.
- The development of the stage motor with flex nozzle by the SMP team, which consisted of Rajaram Nagappa, Dr S.K. Radhamohan, T.T.A. Tharakan, S. Murugesan,

A. Rajamohan, S. Lakshiminarayanan and Kuppuraj among others. As mentioned earlier, the flex nozzle development had been started from scratch with the limited knowledge obtained from literature; the design had several revisions and improvements based on the large number of sub-scale and simulation tests. The rubber formulation required for the elastomeric pads needed very low modulus with adequate shear properties. The team led by K. Adithan took up the challenge and successfully came out with a NATFLEX* compound based on natural rubber. Flex seal was realized using 15CDV6 conical reinforcement shims and elastomeric pad pre-moulded using NATFLEX and bonded through secondary bonding technique. The epoxy-based adhesives called Fusor-305 were imported with suitable primer from Lords Corporation, USA. I am happy that the hard-earned experience described here had given enough confidence for the SMP team to take up further successful flex nozzle developments for motors ranging from 1 m diameter to 3.2 m diameter. Later, the NATFLEX formulation was improved to low modulus compound, LMC-1 and LMC-3, to reduce the actuator load requirement. Flex seal reinforcement shims were realized in spherical configuration to have higher vectoring capability of ±7.8°. The realization process improved from secondary bonding to compression moulding for later HPS3 and S200 motors, thereby enabling faster realization.

- The igniter used was a pyrogen with fast-burning propellant based on the HTPB formulation housed in

---

* Name given by ISRO

a fibre glass/Epoxy case. The RMSA, along with the pyro chain that consisted of ETA (explosive transfer assembly), manifold and TBIs (through bulkhead initiators), was used to initiate the booster charge, which in turn ignited the igniter grain. A similar configuration was used later in the S125-ST01. The igniter hot gases were expelled through six orifices called igniter nozzle inserts to impinge on the main propellant with adequate heat flux to cause normal burning of the motor grain. Here I recall the effort taken by the Igniter and Pyrotechnic Group (IPG) led by V.V. Babu, C.B. Kartha and M. Ravindran in developing an ignition system capable of igniting the motor at sea level as well in vacuum, to suit the flight requirements.

The systematic development route adopted by the solid motor team had not derailed the programme in spite of two consecutive failures during the static testing of the PS3 third and fourth motors. There were frequent review meetings for real-time decisions, which I had the opportunity to chair during every static testing of the PS3 motor (Photo Nos 21 and 22).

The year 1989 was a golden one for ISRO in general and the solid motors community in particular. The first static testing of PS3 ST01 on 26 April 1989 and the first static testing of the S125 booster on 21 October 1989 were grand successes that crossed important milestones for the PSLV programme. The PS3 ST01 had a propellant system with 10 per cent offloading of grain; the bold decision was taken to use the realized motor to demonstrate the TPS performance, especially in the nozzle, wherein there was concern of slender thermal margin.

17. PSLV first stage S125 motor segment assembly: End surfaces inhibited to prevent burning

18. PSLV first stage S125: Integrated 5-segment vertical assembly.

19. PSLV first stage motor S125, at six component horizontal test bed at the static test and evaluation complex (STEX), Sriharikota Range (SHAR) for static test

20. PSLV first stage, integrated five segments, nozzle and core base shroud on the launch pad at Satish Dhavan Space Centre (SDSC)

21. PSLV third stage motor PS3 with handling fixture

22. Static test of PSLV third stage motor PS3 at STEX, SDSC

The nozzle used was of a submerged configuration with a fixed seal and with an area ratio of 8 to have an under-expanded flow in sea-level conditions and to study the thermal environment in the submerged regime, wherein the flex seal was located. The motor performance was as expected and the programme got the green signal. The test of the second PS3 motor – with the flex seal guided in null condition and the propellant fully loaded, without the actuation control system – was carried out successfully on 25 January 1989.

The third and fourth motors were static-tested (ST03 and ST04) with the flight configuration along with the flight version flex nozzle control (FNC) system except the nozzle, which was truncated at area ratio 12.3 to accommodate the actuator and mirror image potentiometer attach points. Computational fluid dynamics (CFD) analysis was carried out by the team led by Dr B.C. Pillai to study the flow pattern at 12.3 area ratio and they cleared the design as over-expansion was expected only at the burning end. Since the FNC system was employed, the actuation trials had to be carried out after assembling the nozzle and the actuation system with the motor prior to taking the motor to the test bed. Hence, the actuation trials were carried out in 2CR of STEX by pressuring the motor to 5 ksc as the vectoring of flex nozzle without chamber pressure would be detrimental to the flex seal.

Both the tests PS3 ST03 (4 May 1990) and ST04 (28 November 1990) ended in failure. The ST03 failure was due to gas entry at the interface between the nozzle end boss and insulation, which pointed to a weakness at the motor/nozzle interface. The failure was at 9.7 seconds after the ignition. I recall that Dr D. Srinivasan, Project Director, PSLV, immediately constituted a committee to investigate the reasons for the test failure – with M.C. Uttam as the chairman, R.V. Perumal, M. Annamalai, N.S.C. Rao, G.K. Kuruvilla, K. Ramachandran and Dr B.C. Pillai as members, and Dr S.K. Radhamohan as the convener. The Overall Review Committee (ORC) chaired by Rajaram Nagappa and a failure analysis committee [FAC] headed by R.V. Perumal reviewed the observations and recommendations of the committee on 12 and 14 May 1990, respectively, within ten days of the setback.

I still remember that the recommendations of the ORC and the FAC were presented to me for a final decision. The observations and data presented clearly indicated the weakness of the nozzle/motor interface and the necessity for improvement. In addition, since there was a joint rotation of nozzle/motor joint to the extent of 0.4° due to chamber pressure and this aspect resulted in a gap opening at the O-ring region, pressure sealing had to be improved with higher-diameter O-ring.

After implementing the recommendations, the PS3 ST04 motor was realized and the static test was carried out on 28 November 1990, which also ended in failure. This time, the redesigned nozzle/motor interface was intact whereas the problem was in the motor/igniter joint region. This incident was of serious concern to the director of the VSSC and the chairman of ISRO.

Considering the serious impact of two consecutive failures, the director of VSSC constituted a high-level FAC on 29 November 1990 with me as the chairman and solid propulsion experts M.R. Kurup, Srinivasa Setty, Rajaram Nagappa, M.C. Uttam, Dr V.N. Krishnamurthy, Dr S. Srinivasan, M. Annamalai, E.V.S. Namboothiri, T.C. Pratapchandran and S.C. Sharma as members, with Dr D. Srinivasan, Project Director, PSLV, and Dr S.K. Radhamohan,

Project Director, SMP, as convenors. The committee inspected the test bed, tested articles and debris along with test data and video clippings and had four sittings to discuss various failure theories and probable reasons with the teams concerned, including System Reliability (SR) groups of the VSSC and SHAR.

Based on the deliberations, we concluded that the inadequacy of bond integrity between the insulation and HE boss was the main reason for the failure. One may ask the relevant question as to why the same situation was not revealed in ST03. It may be noted that a significant change in the ST04 with regard to the ST03 motor was the introduction of one layer of 2 mm thick, room-temperature-cured ROCASIN rubber called RLC. There was a requirement to carry out the acceptance testing of the motor case with nozzle end flange, named bucket flange, to simulate the joint. With the bucket flange assembled with the case, the rubber bladder that was used inside the motor case to avoid water contact could not be used as the bucket flange as its geometry would puncture the rubber bladder. Hence, RLC was instead laid inside the motor case for PPT (Pulse Plasma Thruster), and after test and drying, the case was sent to SPROB for further insulation processing with RLC as the base layer.

Within a month, in January 1991, our committee had submitted the report to the director of the VSSC, with the following recommendations.

- The use of RLC was to be dispensed with and the case had to be tested with flat plate closure instead of the bucket flange, as was followed earlier. For the acceptance of bucket flange, a dedicated flight version motor case had to be identified as a test rig.

- Head boss ID to be covered with integral insulation as a cap to avoid secondary bond line exposed flame front as recommended for nozzle end (NE) boss.
- The FNC system trial actuation test was to be carried out with separate empty motor case for pressurization to 5 ksc instead, using the motor with propellant for pressurization.

With these recommendations, the PS3 ST05 motor was realized and the static test was carried out without the FNC system and by guiding the flex nozzle in the axial direction. The test was successful and the motor performance was as predicted, which placed the programme once again on the groove. Subsequently, the following static tests were carried out to demonstrate the motor performance meeting the various objectives.

- PS3 ST06: With flight configuration along with FNC system, vectoring of the nozzle was carried out.
- PS3 ST07: The flight configuration was without FNC system, and motor placed in the vacuum chamber to demonstrate the vacuum ignition of the motor. The static test programme originally envisaged was to test the motor in the high-altitude test facility, and accordingly, action was taken to establish the facility at STEX. However, there was apprehension regarding starting the diffuser with the motor initial pressure being less than 25 ksc. Experience of the AS4 motor test in HAT (high-altitude test) also did not give enough confidence to proceed with the realization of HAT facility for PS3. On the other hand, what we were sacrificing by not doing the HAT was that

the TPS (Thermal Protection System) performance of nozzle liner downstream of area ratio 12.3 could not be demonstrated. However, the performance of the liner up to 12.3 area ratio could be extrapolated to predict the downstream performance.

- Another objective was the thrust measurement and computation of test Isp. It was observed that the vacuum Isp based on the sea-level data and the theoretical prediction based on the CFD analysis gave acceptable results.
- PS3 ST08 and ST09 were conducted to demonstrate the repeatability in the ballistic performance and to confirm the post-test margins in the various subsystems. Both the tests were employed with the FNC system and the vectoring of the nozzle was carried out with planned duty cycle.

With the nine static tests, the PS3 motor was fully qualified for PSLV. Though the performance of the motor in PSLV-D1 was normal, there was no adequate flight data due to mission failure, which was attributed to a software glitch that disabled the control system. However, the motor performance was excellent and as predicted in PSLV-D2. Subsequently, another static test (ST10) was carried out with optimized TPS in the motor and nozzle. The same design was flown up to PSLV-C3 till the HPS3 motor was inducted in PSLV-C4.

I would like to place on record the effort made by the design and development team of solid propulsion. There was a requirement from a foreign source for the PS3 motor but with improved performance and mass ratio. The mass ratio demanded was similar to SRMs of IUS, with vectoring capability requirement of ±4° in the actuator plane. ISRO approved the development of the High-Performance Motor (HPM) of PS3. Accordingly, the solid motors team, with the help of the SDAs concerned, developed the HPM with the available technology and the materials used in the development of PS3. The design features were as follows:

- Low burn rate propellant with PSOM formulation. (Impact of higher action time of 120 seconds from 80 seconds for PS3 was acceptable for use in the PSLV missions).
- Reduced nozzle throat diameter for the same motor case pressure capability. Reduction in throat diameter (166 mm from 195.4 mm) resulted in smaller size nozzle (mass reduction from 320 kg to 250 kg) with higher area of 69 and Isp increase of 5seconds.
- Flex seal designed was capable of vectoring to ±4°. (Lower shear modulus rubber LMC with 1.8 ksc shear modulus was used with aft pivot point flex seal so that the same actuators could be used for HPM, thereby avoiding the new development.) However, the PSLV mission demand was only ±2°. Hence the flex nozzle design was revised suitably.
- The Kevlar/Epoxy case was optimized with modifying the unequal opening ratio to reduce the inert mass. (Inert mass reduced from 320 kg in PS3 to 285 kg).
- Propellant loading increased by 300 kg.

The propellant mass ratio (that is, the ratio of propellant to total motor) improved from 0.9 to 0.92 in HPM. The HPM was developed with two successful static tests, qualified for PSLV flight and inducted into the PSLV mission from PSLV-C4 onwards.

Other Solid Motors

Solid propulsion technology developments continued and a number of solid motors was developed, qualified and supplied to enhance the performance and to meet the challenging needs of the sounding rockets and launch vehicle programmes of ISRO, as presented below.

Strap-On Motors for PSLV (PS0)

The PS0 motor is derived from the SLV-3 and ASLV projects. It consists of three segmented 15CDV6 motor cases of 1 m diameter loaded with 8.9 tonnes of HTPB-based propellant having an overall length of 10 m with a canted nozzle with cant angle of 9°. The segment hardware interfaces are designed with simple tang-and-groove joint with single face O-ring and assembled with seventy-two 12 mm dia segment holding pins of 15CDV6 material. The nozzle employed was simple convergent-divergent with thermal protection system using ablative liners. Graphite throat insert was used to limit the erosion rate. The area ratio of the nozzle was limited to 6.4 due the constraint in the vehicle interface (Photo Nos 23 and 24).

Out of the six PS0 motors in PSLV, two (one each of ground and air lit) had the provision for SITVC (Secondary Injection Thrust Vector Control) system with 8 kN side thrust capability to augment the roll capability of the vehicle. The ignition system employed was pyrogen type with a cartridge-based ignition train capable of igniting the motor within the specified ignition delay and ignition characteristics.

Subsequently, the capability of the PS0 motor has been enhanced by stretching the segment length from 3 m to 3.4 m and increasing the propellant loading from 9 tonnes to 12.3 tonnes. This change has increased the payload capability of PSLV to the extent of 150 kg. PS0-XL motors were inducted into PSLV from PSLV C-11 (Chandrayaan-1) onwards after a successful static test.

Special Purpose Motors (SPMs)

In addition to the development of stage motors, Special Purpose Motors (SPMs) were required for the separation of spent stages to impart the retro thrust. Moreover, when the sustainer is a liquid stage, SPMs were essential to maintain the ullage volume in the propellant tank to prevent sloshing, and to maintain the acceleration of the vehicle by providing the forward thrust at the time of short coasting between the burnout of the booster motor and the ignition of the sustainer stage. PSLV, GSLV and GSLV-Mk-III have these special purpose motors. All the SPMs, except ullage motors, are basically short-duration motors that burn for 1–2 seconds with the constant thrust demanded for the mission. It was a challenging experience to develop the motor with 90° canted nozzle. Though different types of special purpose motors were developed and qualified for various projects, efforts had been made to have commonalities in the hardware, propellant formulation to avoid new developmental issues.

Special Purpose Motors for Chandrayaan-1

There was a need to develop very small solid motors for the spin, de-spin and de-boosting of the Moon Impact Probe

23. PSLV strap-on motor PS0 with canted nozzle on test bed at STEX, SDSC

24. Static test of PSLV strap-on motor PS0 at STEX, SDSC

25. Spin motor: Special purpose small solid motor for Chandrayaan-1 mission

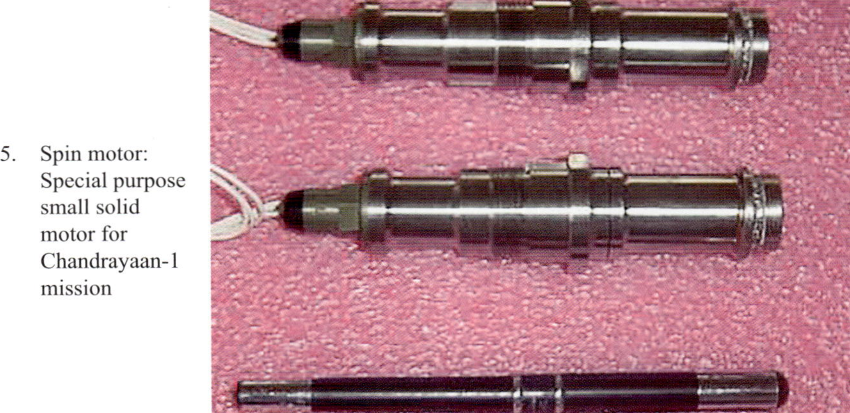

26. De-spin motor: Special purpose small solid motor for Chandrayaan-1 mission

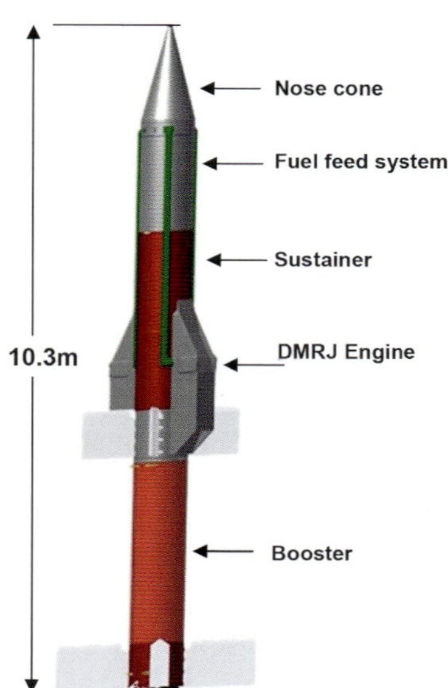

27. RH 560 rocket modified for Dual Mode Ramjet Propulsion-Flight Technology Demonstrator DMRJ-FTD

28. DMRJ-FTD ignited on the launcher at SDSC

(MIP). The spin (Photo No. 25) and de-spin (Photo No. 26) motors were designed with titanium alloy motor case and nozzle without TPS to avoid issues such as degassing and outgassing. An HTPB-based propellant with pyrotechnic igniter was employed. These are the smallest solid motors developed in ISRO.

The size of the motor is comparable to that of a pen. The propellant mass is 8 gm for spin motor and 6 gm for de-spin motor; the corresponding total impulse was 15.5 Ns and 11.5 Ns respectively.

The de-boost motor was designed with Inconel-718 motor case hardware and the insulation material was zirconia (Zirconium dioxide) to avoid the issue of degassing. To avoid the possible impact due to particle impingement on the orbiter, 980 gm of HTPB-based propellant, with only 2 per cent aluminium was used. The total motor mass was around 2.4 kg and the total impulse achieved was 2,250 Ns with the action time of 3 seconds.

Low Burn Rate Solid Booster for Reusable Launch Vehicle Technology Demonstrator (RLVTD)

Towards mastering the technology for the reusable launch vehicle of the future, the Reusable Launch Vehicle-Technology Demonstrator (RLVTD) was designed and developed. The objective of RLV was to take the winged body to an altitude of 30–35 km with a hypersonic velocity (>Mach 6), limiting the dynamic pressure to reduce the aero-load on the structure. This requirement for the solid booster entailed the challenging task of developing a motor with a propellant having low burn rate formulation, thereby resulting in longer burn duration. The RLVTD booster design was carried out with the low burn rate propellant formulation that met the mission and envelope requirements. The motor was of 1 m diameter with three segments based on the pedigree of PS0 motor. The design was qualified through two successful static tests for the RLVTD flight.

Solid Motors in the RSRs and Advance Technology Vehicle

Over the years, ISRO has designed, developed and launched a family of these rockets under the generic name Rohini Sounding Rockets (RSR) to serve a range of scientific missions. They are used to carry out research in areas such as meteorology and upper atmospheric processes up to an altitude of 500 km. The currently operational RSRs are RH-200, RH-300 and RH-560. So far, more than 2,000 sounding rockets have been launched. The RSRs are launched from all the three ISRO ranges – TERLS in Thiruvananthapuram, the Sriharikota launch complex in Andhra Pradesh and the National Test Range (NTR) at Balasore, Odisha. Apart from the scientists of ISRO, those from foreign countries have also been using RSRs to study the neutral and ionized regions of the upper atmosphere, the equatorial electrojet and X-ray astronomy and for meteorological research.

The RSR Family of Solid Motors

The development of the RSRs has not only helped Indian scientists conduct high-altitude experiments but also played a significant role in the realization of SLV-3, India's own satellite launch vehicle.

Apart from the sounding rockets application, the RH series (RH-560-Mk-II) solid motor has been successfully employed in the dual mode ramjet propulsion flight technology demonstrator experiment (Photo Nos 27, 28 and 29).

S200 Strap-On Booster for GSLV-Mk-III

In 2002, the GSLV-Mk-III project was initiated with the mandate to develop a launch vehicle capable of injecting 4,500 kg satellites into Geostationary Transfer Orbit (GTO) and to achieve self-reliance in launching all the communication satellites of the INSAT series. GSLV-Mk-III was configured as a three-stage, simple and reliable launch vehicle with two large strap-on solid boosters loaded with 200+ tonnes of propellant called S200 as the first stage. It was capable of providing a lift-off thrust for the vehicle with 640 tonnes as the take-off mass and to achieve a speed of 2 km/s at burnout at 62 km altitude, along with the operation of the core L110 liquid stage, which is ignited at around T+110 seconds, where T stands for time of ignition of the stage.

S200 Motor Systems: Salient Features

The S200 strap-on booster is a 3.2 m diameter (Photo No. 30) motor that consists of three segments of M250 motor case loaded with 207 tonnes of HTPB-based propellant. The proven nitrile rubber formulation (ROCASIN) is used as motor insulation. The flex nozzle with a vectoring capability up to ±7.8° is employed. The actuation load required is gained by the electrohydraulic actuator (two of them, 90° apart) with 30 tonne (minimum) capacity powered by hydro-pneumatic power pack in blowdown mode. Motor ignition is achieved by the pyrogen ignition system that comprises main and secondary igniters with safe arm, ETA and TBI configuration.

The first static test of the S200 motor with FNC system was carried out successfully on 24 January 2010, and this was a major achievement for the S200 team and the entire solid motor community. It was also a significant milestone for the GSLV-Mk-III project. The second static test was fruitfully conducted on 4 September 2011.

The first experimental mission of GSLV-Mk-III (LVM3X/CARE) with S200 and L110 liquid stage with a dummy C25* was successfully launched on 18 December 2014, thereby validating the flight-worthiness of the S200 motor as a strap-on booster. The mission study, based on the static test performance of the motors in ST01and ST02, indicated the need for design modifications to reduce the dynamic pressure in the transonic range. The ST03 static test was successfully conducted on 14 June 2015 at SDSC, with all the design modifications and improvements. By the successful third static test, all the planned objectives were met and the design was cleared for the GSLV-Mk-III flight. Subsequently, three flights – GSLV-Mk-III-D1 and D2, and M1, the prestigious Chandrayaan-2 mission – have been successfully achieved with the consistent and excellent pair performance of the S200 strap-on boosters. This feat has proved the maturity of the solid propulsion community in the area of reliable design/analysis, and consistency in the processing with stringent quality control and safety measures.

---

* Dummy cryogenic stage with 25 tonnes of propellant.

8

Launch Vehicle Control Systems

THE AUXILIARY PROPULSION SYSTEM Unit (APSU) was engaged in the ongoing activities of (i) the satellite propulsion systems division (SPSD) at Bengaluru, with Dr K. Anantharam as the head; (ii) the Launch Vehicle Control System and Components Division (LCSD) at Valiamala near Thiruvananthapuram (Propulsion Group of the VSSC), with N. Sreedharan Dhas as the head; and (iii) the pressure transducer unit (PTU), with V. Srikantan as the head.

As the programme director of APSU, I organized all development activities related to the control systems of ISRO launch vehicles, and the components for control systems for launch vehicles, satellites and liquid propulsion stages at Valiamala, with N. Sridharan Dhas as the divisional head. This division had three sections: (i) the Reaction Control Systems (RCS) with K. Sivaramakrishnan Nair as the head; (ii) the Secondary Injection Thrust Vector Control (SITVC) system with K. Ramajogi as the head; and (iii) the components division with M.K.G. Nair as the head. The team made significant contributions to the launch vehicle programmes of ISRO. The launch vehicles had also been developed and used the Fin Tip Control (FTC) system and Flex Nozzle Control (FNC) system in the projects. I shall now present the activities of my team.

Reaction Control Systems (RCS): The Background

Dr Vikram Sarabhai gave his approval in 1968 to initiate studies on and develop liquid propulsion systems, which have multiple applications in the Indian Space Programme. Over a period of time, it led to the development and supply of RCS for ISRO's launch vehicles and satellite projects. In 1969, K. Sivaramakrishnan Nair and A. Chandran of the PSN, SSTC, along with their colleagues, designed the first 5 kg thrust, pressure-fed bipropellant liquid engine with hypergolic propellants, aniline and red fuming nitric acid. This development work continued in the PSN, SSTC and, subsequently, in the propulsion group of the VSSC under my leadership.

Initially, a team led by A. Chandran and including S.C. Ghosh, G. Moses, M.A. Haleem, C.G. Balan, D. Anandan and M. Jeyaram started the liquid propulsion system development under the overall responsibility of E.V.S. Namboothiri. When A. Chandran left for a foreign assignment in DOS and Namboothiri joined the SLV project team, the development activities were reorganized under Dr K. Anantharam, with two sections being created. The first one, for bipropellant RCS developments, was under K. Sivaramakrishnan Nair, assisted by C.G. Balan, Raju P. Thomas, P.C. Cherian, V. Raveendranath, P.A. Thangappan Nair and P. Sadasivan Nair.

The second one, for monopropellant RCS developments, consisted of two subgroups – one for hydrazine systems, under the leadership of N.S. Vidyasagar who was supported by A.K. Ray, S. Venkateswaran, K. Muraleedharan and G. Parameswaran, and another for hydrogen peroxide systems under the leadership of R. Palanisamy, assisted by S. Radhakrishnan and R. Krishnan Potti. After some development and evaluation, a decision was taken by me to proceed with hydrazine RCS and to give up the hydrogen peroxide system. Consequently, S. Radhakrishnan joined the monopropellant RCS development and R. Krishnan Potti joined the bipropellant RCS development.

The RCS developments expanded in quality and quantity. When APSU was established, K. Sivaramakrishnan Nair took over the development and supply of RCS to the launch vehicle projects of ISRO. The system design and development teams were ably supported in assembly, integration, test and evaluation by a team of engineers and technical assistants, which included S.S. Jayaprakash, K. Gopinathan Nair, K. Nanukkuttan Nair, S. Jayachandran, C.N. Dhanapalan, R.V. Raman, Sam Zackaria, P. Hariharan, Ajayakumar, S. Kothandan and Ratnakaran.

The RCS, with its on-off capabilities, provides 'forces on command' from the guidance system of launch vehicles to control the pitch, yaw and roll moments of the vehicle and steer it along the required flight path. The bipropellant liquid engines in RCS use hypergolic bipropellants (fuel and oxidizer), which spontaneously react when they come into contact, initiate combustion and produce thrust by the discharge of the combustion products through the nozzle of the engine. Thus, by injecting the flow of the propellants into the engine, one can generate thrust and by stopping the flow, shut down the thrust. These engines are pressure-fed using high-pressure gas bottles with nitrogen or helium or any other inert gas.

The RCS is a self-contained package with liquid engines, propellants storage and feed lines, propellant feed system with high-pressure gas bottles, control components to regulate flows and ensure safety in operation, electrical and instrumentation systems, and mechanical systems to integrate all elements and provide structural integrity to the system. The specifications of these RCS packages are decided by the launch vehicle requirements. These packages are designed, tested, qualified and integrated at appropriate locations in the launch vehicle as per the configuration designs.

Reaction Control System (RCS): Launch Vehicle Projects

The main responsibility of the RCS team was to design, develop and supply RCS packages to SLV-3, ASLV, PSLV and GSLV projects. Moreover, the RCS team worked closely with the launch vehicle teams in integrating these RCS packages with the specified stages of the launch vehicle and also in the pre-launch operations to service fluids to these RCS packages.

During the initial stages, prior to the supply of RCS to the launch vehicle projects, the team developed and supplied RCS packages to the RH-300 and RH-560 control flights, which gave valuable experience to the members.

Monopropellant RCS for RH-300

In order to gain confidence and experience in packaging an RCS system in the very limited space available (usually around the annular space between the main engine nozzle and the inter-stage connecting the two stages or in a restricted space provided between the stages), an RCS system was designed to de-spin the payload attached to the top of the RH-300 sounding rocket at the end of the mission. The space allotted for this RCS was between the top of the second stage and the payload.

The system consisted of four 1 kgf monopropellant thrusters with independent solenoid valves, and a metallic near-spherical propellant tank split along the meridian plane and fastened by bolts. A diaphragm made of butyl rubber separated the pressurizing nitrogen gas from the hydrazine propellant and facilitated the positive expulsion of hydrazine into the de-spinning thrusters. Nitrogen was stored in a toroidal tubular tank at a pressure of 140 bar and was used to pressurize and feed the propellants. The propellant tank pressure was maintained at 20 bar by an electronic pressure regulating system in on-off mode. Stainless steel tubes of 3 mm diameter were utilized for the pressurizing gas line as well as the propellant feed lines.

The thrusters were mounted tangentially to the outer periphery of the shroud of the segment accommodating the RCS package, all in anticlockwise direction (the original spin imparted by the spin rockets was in the clockwise direction). All the four thrusters were called into operation simultaneously to impart the necessary de-spin angular momentum. The regulating system consisted of a pressure transducer to sense the tank pressure, along with a switch to give command to the gas line solenoid valve to open or close depending on the tolerance band of regulated pressure. The catalyst used initially was developed by the Materials and Quality Control Division (MQC) of the VSSC. This RCS system was realized during 1974–75. After two static tests, the RCS de-spin package was successfully flight-tested in October 1975.

Monopropellant RCS for RH-560

Following the successful trial flight of the RH-300 de-spin RCS package, another RCS system with pitch, yaw and roll control was conceived for flight trial in the RH-560 controlled flight in 1975. This RCS had four 25 kgf monopropellant hydrazine thrusters for pitch and yaw control and four 1 kgf hydrazine thrusters for roll control integrated in the annular space available between the second stage (RH-300) nozzle and the inter-stage connecting the second stage to the RH-560 first stage. The thrusters were called into operation by giving command to the respective solenoid valves. The system had one fibre-reinforced plastic (FRP) propellant tank with butyl rubber bladder for the positive expulsion of propellant. Nitrogen gas stored at a pressure of 140 bar was regulated through a single-stage pressure regulator to 20 bar and admitted into the propellant tank. The system had a pyro valve to isolate high-pressure nitrogen from the pressure regulator inlet till the system was initiated in flight, a solenoid-operated vent valve to let out gas during an emergency, a safety pressure relief valve to release pressure in the event of a malfunction of the pressure regulator, and QC/DC (Quick Connect and Disconnect) couplings to fill and drain propellant and gas. The proto-package was static-tested twice on the ground and was refurbished and flight-tested in the one and only controlled flight of the RH-560 from SHAR in early 1978.

Bipropellant RCS Package for RH-560

This RCS package was designed and accommodated in the annular space between RH-300 (second stage) main nozzle and the inter-stage connecting RH-560 to RH-300. The system had two 1.0 litre capacity FRP oxidizer (RFNA) tanks (cylindrical in shape) and two similar FRP tanks for fuel (hydrazine) of the same capacity. The propellants were taken inside the polyethylene bladders and qualified for ten days' storage compatibility to ensure positive expulsion. In addition, there were two FRP gas tanks, each of 750 cc capacity. Nitrogen gas was stored at 150 bar pressure, regulated down to 30 bar pressure and admitted to the propellant tanks for the positive expulsion of propellants to the thrusters. The system had four 50 kgf thrusters for pitch and yaw control and four 10 kgf thrusters for roll control. It had solenoid-operated on-off valves for thruster control, check and vent valves for safety, and sensors for health monitoring in the pretest operations and during testing. The proto-package of the system was realized and tested twice but the flight test did not take place due to non-availability of RH-560 at the time.

Bipropellant RCS for SLV-3-S2 and ASLV-S2

The RCS system for the second stage of SLV-3 (Photo No. 31) was designed for a total deliverable impulse of 11,000 kg-s. This RFNA/hydrazine system had three oxidizer tanks of 7.5 litres capacity each and three fuel tanks of the same capacity. Four 2.5 litre gas tanks, made of FRP material and having a pressure rating of 210 bar, stored nitrogen gas. This high-pressure nitrogen was regulated through a single-stage pressure regulator and admitted into the propellant tanks for positive expulsion.

The initial proposal was to make these tanks from FRP material and to use the bladders from polyethylene material as in the RH-560 RCS package for positive expulsion. However, a catastrophe that occurred during the static test of the package forced the development team to change the tank material to 304 grade stainless steel (SS 304) and the bladder material from polyethylene to the more compatible Teflon FEP (fluorinated ethylene propylene) material. This RCS had four 250 kgf thrusters for pitch and yaw control during powered phase, four 50 kgf thrusters for pitch and yaw control during coasting phase, and four 10 kgf thrusters for roll control during the entire second stage operation. Each thruster operation was controlled by two independent on-off solenoid valves that admitted the propellants into the respective thrusters.

This system had safety vent valves, safety relief valves and check valves to ensure safety during operation. In addition, various pressure sensors and temperature sensors were mounted at appropriate locations for health monitoring during the pre-launch as well as in-flight operation. After the second successful flight of SLV-3, the 50 kgf thrusters were removed from the RCS system because there was very minimal requirement of pitch and yaw correction during the coasting phase, as observed during flights. The same RCS system was used for the second stage control of ASLV too.

Monopropellant RCS for the Third Stage of SLV-3 and ASLV

The RCS for the third stage of SLV-3 (Photo No. 32) uses hydrazine as propellant. This system has four 25 kgf thrusters for pitch and yaw control during the powered phase of the third stage operation, and four 5 kgf thrusters for pitch and yaw control during the coasting phase of the third stage operation. In addition, four 1 kgf thrusters were used for roll control during the entire third stage operation. In this system, hydrazine was stored in two FRP propellant tanks, each with a capacity of 10 litres. EPDM bladders were used for positive expulsion. High-pressure nitrogen gas was stored in the FRP tank at 210 bar pressure and regulated down to 35 bar pressure through a single-stage pressure regulator. The same RCS package was used for control during the third stage operation of ASLVs.

RCS for PSLV-S1

After the successful development and supply of RCS for SLV-3s and ASLVs, the development of RCS for PSLVs and, subsequently, for GSLVs was taken up. For the first stage of PSLVs, two identical RCS packages were developed for roll control. Each package had a 600 kgf (at sea level) roll control thruster that could swivel ±45°. This thruster was regenerative-cum-film-cooled with an area ratio* of 4. It used nitrogen tetroxide (N_2O_4) and monomethylhydrazine (MMH) propellant combination at a mixture ratio of 1:4. It also utilized two ablative-cooled 150 kgf (at sea level) auxiliary control system thrusters for pitch control during the tail off of PS1.

The propellants were stored in two spherical stainless steel tanks. Initially two tanks of 125 litres each were used, which were later modified to 163 litres for fuel and 142 litres for oxidizer. The propellants were fed to the thrust chamber

---

* Area ratio is the ratio of the cross section to throat area.

at a constant pressure of 25 bar using stored inert gas for pressurizing, stored at 330 bar pressure in a 34-litre titanium gas bottle. The inert gas was supplied to the propellant tank through a dual port pyro valve, pressure regulator and check valves. These packages were integrated on a separate cylindrical structure and then attached to the main stage. After successful tests and qualification in two proto-packages, these RCS packages were used in PSLV flights.

The system was used for proportional roll control with an electromechanical actuator to swivel the engine during the PS1 thrust phase and three-axis control during auxiliary control phase. This development was a stepping stone towards adopting proportional RCS in launch vehicle systems. Here, I must appreciate the sincere contributions of S. Radhakrishnan, V. Krishnanandan, P. Raveendran, Mahesh Narayanan and P. Gilbert, from its development till flight.

Control System for PSLV-S2

The second stage of the PSLV was a liquid stage. The main engine (Vikas engine) of the stage was gimballed ±4° for pitch and yaw control. Two 35 kgf hot gas thrusters were used for roll control. These thrusters used the hot gas tapped from the main engine gas generator.

Control System for PSLV-S3

The third stage of the PSLV had a solid motor. The pitch and yaw control used the flex nozzle of the main motor, with electromechanical actuators with ±4° angular movement capability. The roll control was by the RCS module with six 5 kgf bipropellant thrusters, integrated in the fourth stage.

Control System for PSLV-S4

The fourth stage of the PSLV was also a liquid stage with twin engines. The pitch, yaw and roll during the thrusting phase of the flight were controlled by gimballing these twin engines. During the coasting phase, the pitch, yaw and roll controls were managed by RCS with six 5 kgf bipropellant thrusters, which utilized the propellant from the main tank. The same RCS was also used for the roll control of the third stage as well as for the precise velocity trimming of the fourth stage before injection to orbit.

Control System for GSLV-Mk-II

Four L40 liquid stages and a solid motor were used in the first stage of the GSLV-Mk-II. The pitch, yaw and roll control of this stage was done by gimballing the four L40 engines, using electro-hydraulic actuators. The second stage of GSLV-Mk-II used a single Vikas engine. The pitch and yaw control was handled by gimballing the main engine and roll control was managed by two 35 kgf hot gas motors as in the second stage of PSLVs. The third stage of GSLV-Mk-II used the Cryogenic Upper Stage (CUS) engine along with two steering engines. These steering engines of 200 kgf thrust were used for the pitch, yaw and roll control of the stage. Electromechanical actuators were used to gimbal these steering engines ±40° in two planes.

Control System for GSLV-Mk-III

A liquid stage L110 with two Vikas engines and a solid booster were used in the first stage of the GSLV-Mk-III. The solid booster with flex nozzle and gimballed Vikas engine were used for the pitch, yaw and roll control of this stage. The upper stage was a cryogenic propulsion system, which used the CE20 engine. The pitch and yaw control of this stage was controlled by gimballing the CE20 engine and the roll control was handled by cold gas hydrogen thrusters, which used the hydrogen from the main propulsion system.

Control System for Reusable Launch Vehicle Technology Demonstrator (RLVTD)

Apart from the development of launch vehicles, a major initiative was undertaken by ISRO to develop and flight-test a reusable launch vehicle technology demonstrator (RLVTD). After detailed studies of the system requirement and available technologies, it was decided to develop a hydrazine monopropellant RCS for this test vehicle in association with Indian industries. The system utilized twelve 200N thrusters, developed at Liquid Propulsion Systems Centre (LPSC), Bengaluru, for the three-axis control of the technology demonstrator vehicle (TDV). Two propellant tanks, each with a capacity of 40 litres, stored the propellant hydrazine. Carbon-overlapped titanium alloy liner gas bottle was designed and developed to store nitrogen gas at a high pressure of 125 bar.

The indigenous development of the catalyst in collaboration with Indian industries was a major challenge. Ogene Systems India, a company in the private sector, successfully developed and qualified an iridium-based catalyst for this project in close association with LPSC. The firm also supplied the required quantity of the catalyst of specified quality for the project on time. Yet another concern was the development of the EPDM rubber-based diaphragm by Vajra Rubbers,

Irinjalakuda, Thrissur district, Kerala for the propellant management system to meet critical mission requirements. The propellant tank forgings were developed by Steel & Industrial Forgings Limited (SIFL), Thrissur, and the tanks were realized at BrahMos Aerospace Trivandrum Limited (BATL), Thiruvananthapuram. This RCS was designed, developed and successfully flight-tested in the maiden launch of RLVTD on 23 May 2016. I congratulate V. Krishnanandan, G. Makesh, K. Shambayya, Mahesh Narayan, A. Manimaran and their team on their vital contributions to the project.

Control System for Space Capsule Recovery Experiment (SCRE)

Space capsule recovery was yet another major initiative taken by ISRO, in 2001, for future manned missions and payload recoveries. Towards this objective, LPSC had developed and supplied a pressurization system module for flotation and RCS. This fast-acting gas pressurization system, with gaseous nitrogen (GN_2) as the pressurizing gas, filled 600 litres of the float within 1.2 seconds to maintain the recovered capsule 'floating'. The development and validation of the system were carried out at the Naval Science and Technological Laboratory (NSTL), Visakhapatnam, and the SDSC. In addition, twelve 22N thrusters, using MMH and MON-3 as propellants, were used for the RCS propulsion and control. The system had two propellant tanks, each with a capacity of 30 litres, which had a propellant management device (PMD) to supply gas-free propellants. SRE-1, a 555kg capsule, was launched on 10 January 2007 using the PSLV-C7, which was ejected at an altitude of 540 km and was subsequently successfully recovered from the Bay of Bengal near the SHAR coast. The efforts of B.K. Venkataramu, S. Venkateswaran, V. Krishnandan, H.S. Venkatesh and their team were instrumental in making the project a success and were well appreciated.

Unfortunate Accident: Bipropellant RCS Development for SLV-3

The development of a bipropellant RCS package for the second stage of SLV-3 was undertaken in 1973 using the RFNA/hydrazine propellant combination. In addition to the ablative-cooled 250 kgf, 50 kgf and 10 kgf thrusters, the system had four fibre glass gas bottles of 3 litre capacity each, three oxidizer tanks of 7.5 litre capacity each and three fuel tanks each with a capacity of 7.5 liters, along with pressure regulator, safety valves and allied plumbing – all integrated on the split inter-stage structure occupying the annular space surrounding the main nozzle of S2.

The RCS team had an excellent record on safety in handling and operating this system with hazardous materials. However, a single accident occurred during the first integrated system static, where some of our engineers and technicians – K. Sivaramakrishnan Nair, P C. Cherian, Raju P. Thomas, L. Muthu, A.M. Hanifa, C.N. Dhanapalan and P. K.N. Nambiar – sustained serious acid burns. After a detailed analysis, it was found that the mishap was due to inadequate coordination between the PSN and SLV teams in carrying out test operations, specifically in simulating the operations in flight using the telemetry system supplied by the SLV.

Initially, all the propellant tanks were made of FRP material with ROCASIN rubber lining in order to arrest the leakage of pressurized gas through the FRP. The process of making the tank involved, (1) making a sand mandrel, (2) laying the ROCASIN rubber lining; and (3) winding FRP on the lined mandrel and curing it in the oven. The sand mandrel is removed later by dissolving the binding material of the mandrel in water, thus disintegrating the sand particles and removing them through the end openings. The failure analysis revealed that one of the bladders in the propellant tank was ruptured due to leftover sand particles inside the propellant tank from the sand mandrel, which was used in making the tank. Subsequently, the raw material of the propellant tank was changed from FRP to stainless steel.

The setback occurred due to an explosion in one of the oxidizer tanks used in the system. After charging the propellants (RFNA and hydrazine) into the package from the mobile propellant charging units, the test team was given clearance to proceed with the boost charging of the pressurizing gas (nitrogen). But due to some mechanical problem in the booster pump, there was a delay of a few hours in rectifying the issue and resuming the boosting operation. When the gas boosting was resumed, a minor RFNA leak was noticed in one of the 10 kg oxidizer feed lines. As per safety protocol, one is allowed to work on the system only when it is not under pressure. The testing team contacted the test director at the control centre, who verified the situation, found that the pressure indications in the control panel of the propellant tanks were ambient, and authorized operations to tighten the ferrule in the feed line and arrest the leak.

In the process, one of the FRP acid tanks exploded, spilling the acid all around and seriously injuring some of the personnel. Obviously, the system was under pressure and the indication in the control panel was wrong. On verification, it was found that on the instructions of the SLV project leader, in order to save the battery life in the telemetry system, the SLV technician who operated the telemetry system had stopped the acquisition of pressure data from the propellant tanks and feeding it to the control centre, without informing the PSN test director. The gap in coordination between the teams was soon bridged.

The Development History of the Low-Thrust Liquid Engines by RCS Team

Over the years, R&D was carried out in different areas of liquid engine development, including injectors, thrust chambers, nozzles, propellants, feed systems and so on. Several types of liquid engines were developed, as summarized below:

i) This liquid engine of 5 kgf thrust was the first pressure-fed, short-burning, bipropellant thruster, which used commercially available aniline as fuel and RFNA as oxidizer, with no solenoid valves for flow control, and had an uncooled stainless steel thrust chamber. It was designed and manufactured using in-house resources (facilities and manpower), assembled and tested in a primitive test stand with minimal instrumentation facilities at Valiamala, Thiruvananthapuram. This engine was demonstrated successfully before Dr Vikram Sarabhai in September 1969. Subsequently, the test operations were shifted to the Thumba beach. The thrust of this engine was uprated to 10 kgf. Flow control components, including solenoid valves, pressure regulators, vent valves and so on were designed, developed and incorporated in the 10 kgf thrusters. The on-off capability of the thrusters was demonstrated, and their potential for the development of reaction control systems for launch vehicles and satellites control systems was established.

ii) For this 200 kgf general purpose liquid engine (GPLE), different injector configurations were studied, including multi-element impinging doublet, triplet unlike impinging, single-element coaxial sheet impinging with a continuous sheet type and discrete oxidizer sheet type. Many kinds of propellants were also experimented with, as outlined below.

   a) Oxidizer: RFNA with varying NO_2 content between 6 per cent and 16 per cent, and with various additives such as sulphuric acid, ferric chloride and potassium permanganate.

   b) Fuel: Aniline with various percentages of furfuryl alcohol as additive, hydrazine and UDMH.

   c) Thrust chamber: Silica phenolic liners with different composition were studied for chamber cooling. Nozzle throat inserts made of molybdenum-copper or tungsten-copper were studied in order to develop thrusters for the RCS of SLV-3 and ASLV launch vehicles.

iii) Bipropellant liquid engines of different thrust levels were designed, developed and supplied for use in SLV-3 and ASLV launch vehicles, which included 10 kgf, 30 kgf, 50 kgf, 200 kgf and 250 kgf ones. The 250 kgf and 50 kgf thrusters were redesigned with 90° cant angle nozzle for the SLV-3 second stage control system, using RFNA and hydrazine as propellants, qualified on the ground and flight-tested on SLV-3 and ASLV flights.

iv) A 50 kgf thrust, long burning (more than 500 seconds) liquid engine had an oxidizer-cooled, regenerative stainless steel thrust chamber, using single-element coaxial swirl injector and RFNA/hydrazine propellants with a mixture ratio* of 1.65. It was developed and used to demonstrate drilling operations by rocket exhaust in the Rocket Drill project.

v) A 30 kgf/10 kgf dual thrust engine for possible use in RCS application, with RFNA and hydrazine as propellants was developed.

vi) A 600 kgf thrust, ablative-cooled, liquid engine using RFNA as the oxidizer and aniline with furfuryl alcohol as fuel, which burned for 20 seconds, was developed and supplied for the first liquid stage of ISRO. It was flight-tested from Sriharikota in 1973 and 1974.

vii) A 3,000 kgf thrust, ablative-cooled, liquid engine using RFNA as the oxidizer and aniline with furfuryl alcohol as fuel was developed and tested at a newly established Engine and Stage test stand at Sriharikota in 1976. It was supplied to develop a liquid stage for flight test as a second stage of RH-560 in 1977 and as a single stage in 1978.

viii) A 3,000 kgf doublet radial injector engine, using RFNA as the oxidizer and UDMH as fuel and silica phenolic ablative-cooled thrust chamber was designed

---

* RFNA is the oxidizer and hydrazine the fuel. The mixture ratio refers to ratio of flow rates of oxidizer to fuel (O/F).

and manufactured in-house. It was tested twice in a horizontal test bench at TERLS during 1973–74.

ix) A 200 kgf thrust level engine, regenerative-cooled and using coaxial sheet impinging injector (discrete oxidizer sheet), and RFNA as the oxidizer and MMH as fuel was developed and tested successfully for a duration of more than 60 seconds. MMH was used as the coolant in the regenerative coolant passage. Thus, the regenerative cooling of the thrust chamber with fuel as coolant was successfully demonstrated.

x) A 600 kgf liquid engine with N_2O_4 as the oxidizer and MMH as fuel and with a mixture ratio of 1.65 was developed for the roll control thrusters of the PSLV first stage. This engine used single element coaxial sheet impinging injector (discrete oxidizer sheet) and regenerative-cooled thrust chamber and demonstrated a C*[*] efficiency of 92 per cent.

xi) An over 600 kgf engine, developed for the roll control thruster, was upgraded for the fourth stage application of PSLV. When this engine was upgraded for C* efficiency more than 97 per cent, a multi-element triplet injector was designed. The regenerative cooling passage had to be redesigned to prevent burn-through that occurred in the earlier simple design. Consequently, the regenerative cooling was limited to an area ratio of 4.8, and a nozzle extension from the area ratio of 4.8 to 60 was made of columbium

---

* C-star is a measure of the combustion performance of a rocket engine independent of nozzle performance, and is used to compare different propellants and propulsion systems.

C103 material with silicide coating in order to prevent oxidation. The Isp of this updated engine with an area ratio of 60 was more than 305 seconds. This engine was demonstrated with a burning time of over 1,000 seconds with eight restarts.

xii) A 5 kgf thruster was developed for RCS application in pulse mode in PSLV, using N_2O_4 and MMH as propellants. This engine had a single-element coaxial swirl injector with a film-cum-radiation-cooled thrust chamber made of Nimonic 90 material with an area ratio of 40, operating in pulse mode with a cumulative burn duration of 75 seconds. This engine was used in PSLV-PS3 (third stage) roll control as well as for PSLV-PS4 (fourth stage) pitch, yaw and roll control during pre-coasting as well as post-coasting periods.

xiii) The 22N attitude and orbit control system (AOCS) engine was developed for the INSAT programme. This engine had single-element coaxial swirl injector with inner fuel swirler and outer oxidizer swirler. N_2O_4 was the oxidizer and MMH was the fuel, operating with a mixture ratio of 1.65 and delivering an Isp of 290 seconds. It had a fully film-cum-radiation-cooled thrust chamber having area ratio 60 with columbium C103 material with silicide coating to prevent oxidation. It was developed and qualified in 1992.

xiv) The 440N liquid apogee motor (LAM) engine with 440N thrust was developed for the INSAT programme, using N_2O_4/MMH propellants with a mixture ratio of 1.60. It had a coaxial swirl injector, with the oxidizer and fuel flowing through the outer and inner flow passages respectively. The radiation-

> cooled thrust chamber was made up of silicide-coated columbium alloy. Mark-I LAM was designed initially with an injection pressure of 15.5 bar. The spacecraft system pressure drop evaluation arrived at an injection pressure of 14.5 bar. The extra pressure drop was adjusted by increasing the radial gap in the swirl flow passages, and the engine was qualified and used in INSAT-2A, 2B and 2C. Owing to this adjustment, the Isp came down to 307 seconds against the specification of 310 seconds (minimum). The design was modified to meet the revised injection pressure of 14.5 bar. With the revised design Mark-II, one engine (D3) was qualified in High-Altitude Test (HAT) for a total burn time of 34,000 seconds with more than forty restarts. This version started flying from INSAT-2D onwards.

To summarize, the challenge was to design, develop and qualify an RCS bearing in mind the space limitation in the launch vehicle. In the development of the thrusters for RCS, the successful development of the ablative throat was a major task. During the initial development, asbestos phenolic ablative liners with graphite throat were used. The high chamber pressure and canted nozzle configuration caused heavy erosion of the throat. Different throat inserts with varied graphite materials and ceramic coatings were experimented with but the attempts failed in the initial development phase. After quite a few trials, molybdenum impregnated with copper and a thin layer of silicide coating was selected for throat inset, and silica phenolic for ablative liner was successfully developed after around 200 tests in more than thirty different configurations. The enthusiastic and steadfast efforts of Samuel Jeyaprakash, K. Gopinathan

Nair, R.V. Raman, S. Kothandan and Ajayakumar deserve special mention in this regard.

SITVC System

The SITVC system is used in launch vehicles to generate large side thrusts in the pitch and yaw directions and to steer the vehicle in flight. When a fluid is injected in the divergent section of a fully flowing nozzle of a solid motor, it triggers a chemical reaction with the exhaust gases and creates a shock wave in the supersonic flow region of the divergent nozzle. This shock wave propagates outward from the nozzle's inner wall. Consequently, the flow through the nozzle is modified and a side thrust is generated.

Development activities related to SITVC for launch vehicles were initiated in the propulsion group under my guidance in 1973 by N. Sridharan Dhas who was the head of the TVC section. Dhas joined PSN from the CGI of the VSSC, along with his engineers K. Radhakrishnan, J. Watson, S. Palaniappan and H. Venkateswara Iyer. Subsequently, more engineers K. Ramjogi, K. Sasidharan Nair, M. Ganesan, M.C. Indira and F.P. Joshua and technicians such as Nagappan Asari, C. David and N. Srinivasan joined the team.

In the initial stages of development of the SITVC system, a considerable number of ground tests were conducted with small solid propellant motors used in RH-125, D-200 and RH-560 by varying the injectant parameters, injection point, injection angle and injection pressure to optimize the system design. Based on the results of these tests and after a detailed literature survey, the aqueous solution of strontium perchlorate was selected by the SITVC team as the injectant. After successful ground tests, the first SITVC system with the strontium perchlorate injectant and a blowdown pressurization system was successfully flight-tested in the RH-560 controlled flight, using an imported single-pintle injection valve.

SITVC SLV3-S1

For the first stage (S1) of SLV-3 (Photo No. 33), the project proposed a SITVC for the pitch and yaw control during the initial 17.5 seconds of flight, and a fin tip control (FTC) system for the remaining part of it since FTC is not effective at lower velocities. Further, SITVC was to be in backup mode from 17.5 seconds onwards till burnout. The side thrust specified for pitch and yaw control was 1,500 kgf. Accordingly, a SITVC system with four three-pintle proportional electro-hydraulic injection valves was designed for S1 of SLV-3. These valves were imported from Moog Inc., USA.

A six-component test stand with the capability to measure pitch, yaw and roll moments, along with the main thrust of the solid propellant motor, was established in one of the STEX test beds to evaluate the performance of the SITVC system. A ground service system with the accessories required was built to conduct the static test. The static test of the S1 SITVC system in the first stage motor of SLV-3 was conducted at STEX in a pre-planned test sequence, designed to meet the requirements of maximum side thrust, simultaneous injection in pitch and yaw ports, and to accommodate the flight duty cycle with a factor of two. The static test was successful and the system performance satisfactorily met all qualification requirements for flight.

The SITVC flight system was positioned between the base shroud and nozzle of the S1 motor. The system used six FRP tanks with ROCASIN rubber bladders to store the injectant and another FRP tank to store hydraulic oil MIL-H-5606 for the actuation of injection valves. Three high-pressure (210 bar) FRP gas bottles were used to store nitrogen gas. High-pressure gas was regulated to 50 bar using two spring-loaded pressure regulators. A pyro valve isolated the pressure regulators from the high-pressure gas bottles. The injectant from the tanks was distributed to four injection valves through a toroidal manifold and four flexible hoses. Burst diaphragms were used in these lines to isolate the injection valves from the injectant. During the development phase, severe shock was experienced whenever the pyro valve was ignited to start the pressurization of the injectant, thereby resulting in gas leakage through the joints in the high-pressure gas line. As a corrective action, the pyro valve was replaced by a slow-opening system start valve (SSV).

The injection sequence was finalized for the static test, taking into consideration the maximum flight duty cycle. During the initial stages, galvanic corrosion was observed in the burst diaphragms made of aluminium. It was solved by bonding Teflon sheet over the surface of the diaphragm and qualifying it with a number of tests. Similarly, in order to overcome the failures experienced in PVC-tube-based flexible hoses, a special flexible hose was developed with Teflon tube that featured end fittings with metal-to-metal joint combined with Viton 'O' ring. The hose was qualified successfully by conducting various tests, including the vibration test.

The SITVC flight system was integrated with S1 of the SLV-3 E01 flight at SHAR and functional tests were conducted successfully. Pre-launch operations such as injectant and oil filling as well as nitrogen gas charging were

29. DMRJ-FTD in flight over SDSC Range

30. S 200: Biggest solid motor of ISRO with 3.2 metre diameter and 200 tonnes of solid propellant and flex nozzle control system, used in pair as strap-on in GSLV-MK-III is assembled for static test on the test bed at STEX, SDSC

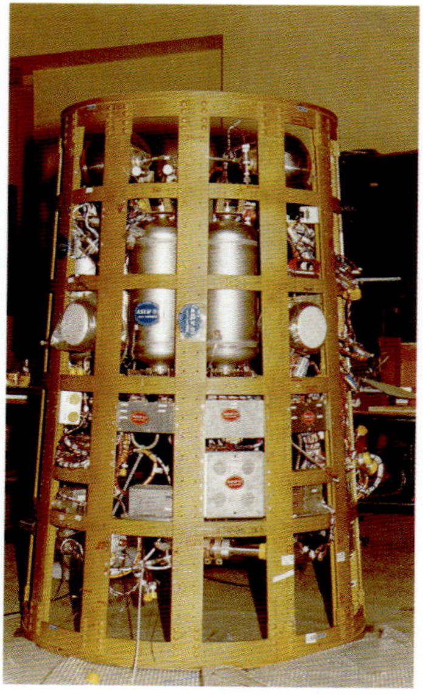

31. Bi-propellant Reaction Control System (RCS) for the second stage of SLV 3 and ASLV

32. Mono-propellant Reaction Control System (RCS) for the third stage of SLV3 and ASLV

33. Secondary Injection Thrust Vector Control (SITVC) system for the first stage of SLV 3 and ASLV

34. Secondary Injection Thrust Vector Control (SITVC) system for ASLV strap-on motors

completed. During the first SLV-3 launch on 10 August 1979, the first stage with SITVC system performed normally, but the mission was a failure due to the malfunctioning of the second stage control system. The second SITVC flight system for SLV-3 E02 was prepared with top priority accorded to it, and was integrated with the S1 motor on schedule, followed by successful pre-launch operations. The SLV-3 E02 launch on 18 July 1980 was successful, and the performance of the SITVC system was normal.

SITVC-ASLV

The ASLV configuration was practically the same as that of SLV-3, except for the two additional strap-on motors identical to those of the SLV-3 first stage (Photo No. 34). The SITVC system was selected to ensure pitch, yaw and roll control for strap-on motors and pitch and yaw control for the first stage. An external stainless steel tank was used to store the injectant and 15CDV6 tank stored the pressurizing nitrogen gas. A high-flow dome-loaded pressure regulator was also designed and qualified for use in the high-pressure gas line. A newly designed injectant start valve (ISV) was used instead of the burst diaphragm for injectant isolation purpose. A bypass solenoid valve was introduced across the ISV to avoid the water hammer effect.

The quantum of work for the ASLV was about thrice that of the SLV-3, and so a separate integration lab was established with minimal facilities. To validate the system parameters, a static test with a strap-on motor that had a canted nozzle was conducted in the six-component test stand successfully. Even as the activities towards the ASLV-SITVC flight system were in progress, the PSLV project was approved. To meet the increased workload, more people joined the SITVC team, such as T.M.T. Nicolas, R. Srinivasan and K.V. Sivaraman for ASLV, and K. Shaji and Jose Paul for PSLV.

In the initial stages of the ASLV programme, imported three-pintle electro-hydraulic injection valves were used. However, since twelve injection valves were needed for each flight, it was decided to indigenously develop and manufacture them. As a result of focused development efforts, these injection valves were developed in record time, and successfully put through a series of qualification tests, including flow test, vibration test and so on. The performance of the indigenously developed injection valves was on par with the imported ones.

SITVC systems for the ASLV-D1 flight were realized and integrated with the first stage (AS1) and strap-on motors (AS0*) at SHAR. After successful pre-launch operations, they were launched in ASLV-D1 on 24 March 1987. Both AS1 and AS0 SITVC systems worked satisfactorily during flight as per command, but the ASLV-D1 mission was a failure due to the malfunctioning of the upper stage. The post-flight analysis showed that the failure was not due to SITVC systems. Subsequently, SITVC systems were realized and supplied for ASLV-D2, D3 and D4 flights. The ASLV-D2, launched on 13 June 1988, was also a failure, in spite of the normal performance of the SITVC systems. The SITVC systems performed satisfactorily and as per command in the successful launches of ASLV-D3 on 22 May 1992, and ASLV-D4 on 5 May 1994.

---

* 0 (zero) stands for strap-ons. The solid motors are designated with PS followed by the number referring to the stage. 0 means strap-ons. Thus it is PSLV solid motor strap-on.

SITVC-PSLV PS1

The PSLV core stage PS1 (Photo No. 35) demanded a huge side thrust of 20,000 kgf, which required very high flow rates of the injectant through the injection valves. The available injection valves used in SLV-3 and ASLV could not meet the large flow requirements. Hence, the PS1 SITVC system was designed with multi-port injection scheme employing twenty-four single-pintle electromechanical injection valves, equi-spaced around the motor nozzle, to meet the high flow requirement of injectant. Moreover, the multiport injection concept promised 20–30 per cent increase in injection efficiency. A single-pintle injection valve, actuated by a DC torque motor was designed, developed and qualified for flight after stipulated tests.

Parallelly, a sub-scale version of this injection valve was developed and qualified. A scaled-down version of the PS1 SITVC system with twenty-four sub-scale single-pintle injection valves mounted on an ASLV core motor was static-tested successfully at the STEX six component test stand. The performance met all the requirements and encouraged the TVC team to proceed with the PS1 SITVC system development. Subsequently, the PS1 SITVC system with twenty-four single-pintle electromechanical injection valves was successfully tested in the PSLV core motor in the newly established six component test stand at STEX.

The PS1 SITVC system used two external 15CDV6 tanks with aluminium oxide coating inside the tank to be compatible with the strontium perchlorate injectant. The injectant was pressurized in blowdown mode. Two injection start valves (ISV), one per tank, were used to isolate the injection valves from the injectant during the initial periods of the launch. All the twenty-four injection valves were connected to a toroidal manifold through twenty-four metallic flexible hoses. The trial integration of the PS1 SITVC flight system was done first at the Valiamala Complex and the final integration of the system with flight motor was done at SHAR.

SITVC-PSLV PS0

While the development of SITVC for PS1 was progressing, the PSLV project came with a new requirement of SITVC system for PS0, a little late in the PSLV programme, with a view to augmenting the roll control capability. The PS0 SITVC system was to be mounted on diametrically opposite strap-on motors. Since the vehicle configuration had already been finalized, the stainless steel injectant tank and all the control components, including pressure regulator, had to be housed in the small space available inside the nose cone of PS0. Four high-pressure FRP gas tanks were accommodated in the space between the nozzle and the base shroud. Two units of the already qualified sub-scale single-pintle electromechanical injection valves were mounted on the canted nozzle divergent of both the strap-on motors. A static test was successfully conducted in the six component test stand at STEX to validate the design of the PS0 SITVC system.

Subsequently, flight systems of PS0 SITVC for the PSLV-D1 mission were realized and the trial integration was done on a dummy motor at Valiamala. The final integration was carried out on the launch pad at SHAR. Prior to the launch of PSLV-D1 on 20 September 1993, all pre-launch operations of PS1 and PS0 SITVC systems were completed successfully. However, during the automatic launch sequence, a relief valve opened accidentally and resulted in the drop of high pressure gas bottle pressure from 55 bar to 50 bar (reseal pressure of relief valve). Thereafter, the pressure remained constant at 50 bar. After a debate, the launch was cleared based on the analysis that blowdown mode pressurization from 50 bar was sufficient for the mission, and the vehicle took off.

All SITVC systems worked satisfactorily during the flight, but the mission was unsuccessful due to the malfunctioning of the guidance system of the vehicle. Based on the satisfactory performance during the PSLV-D1 flight, the PS0 SITVC system configuration was simplified by incorporating blowdown mode pressurization for the PSLV-D2 mission, thereby eliminating the need for pressure regulator and relief valve. Thereafter, configurations of PS1 and PS0 SITVC systems have remained unchanged till date, with a record of excellent performance in all subsequent PSLV missions.

9

Satellite Propulsion Systems

AS THE PROGRAMME DIRECTOR of the Auxiliary Propulsion Systems Unit (APSU), one of my jobs was to organize the satellite propulsion systems developments at the Satellite Propulsion Systems Division (SPSD) of APSU in Bengaluru, with Dr K. Anantharam as the divisional head. This division was co-located with PTU, also of APSU, on the same campus. The SPSD worked closely with the ISRO Satellite Centre (ISAC), now renamed U.R. Rao Satellite Centre (URSC), in Bengaluru, and the launch vehicle control system division of APSU at Valiamala.

Satellite Propulsion Systems Division (SPSD)

The SPSD was responsible for the development of monopropellant propulsion systems for the IRS (Photo No. 36) series and bipropellant propulsion systems for the INSAT (Photo No. 37) series of satellites. Subsequently, the team made excellent contributions to ISRO programmes. As is well known, the satellite propulsion system has two major functions – the first, to raise/transfer the satellite from the orbit in space, where the launch vehicle would have placed it, and the second, to maintain the satellite in the desired attitude and orbit as per the mission requirements. Obviously, the thrust to be produced by the satellite propulsion systems is far smaller when compared to that of the launch vehicle propulsion systems. During my tenure at ISRO, the development of satellite propulsion systems was based on chemical propulsion. I understand that ISRO subsequently initiated developments in ion propulsion and electrical propulsion.

At the outset, I would like to congratulate and thank my colleagues Dr K. Anantharam and K.L. Valliappan for their leadership, and their team members for the excellent work. I also wish to place on record my gratitude to all my senior colleagues at APSU, B.K. Venkataramu, B.R. Rao, S. Mohan, M.N. Reddy, Anantharama Reddy, V. Sivagopal, C.S. Prasad and S. Ramesh and their associates. All of them contributed significantly to the growth of APSU.

Monopropellant Propulsion System for Satellites

The major developments towards monopropellant propulsion systems for satellites were micro-thrusters and their associated elements such as flow control valves, fill/drain valves, fuel tanks and sensors. Developments during those days were challenging in the sense that one had to do everything from drawings and component assembly to performance testing. The initial developments were 1N (Photo No. 38) and 11N monopropellant thrusters (Photo

No. 39) and associated components. In addition, fuel tanks with expulsion diaphragms and surface tension devices were developed to ensure positive propellant discharge throughout the operation (Photo No. 40).

All qualification tests for the first monopropellant RCS and its elements were successfully completed by the end of 1985. The first spacecraft to carry indigenous RCS was IRS-1A, launched by the Russian launcher VOSTOK on 17 March 1988. It provided valuable information on the performance of RCS in flight. The IRS-1A system had sixteen 1N thrusters and associated control components and four 30-litre fuel tanks equipped with specially developed EPR rubber expulsion diaphragms. Similar systems were also flown on IRS-1B and IRS-1C. These spacecrafts outlived their design lifetime of three years and provided valuable data on the orbital performance of RCS. However, the propellant tanks with rubber diaphragm used in these three flights contaminated the fuel (hydrazine), thereby resulting in minor degradation in the thruster performance. Hence, we decided to replace tank expulsion devices with passive, metallic capillary acquisition systems and to get rid of rubber diaphragms.

In the course of the development of the bipropellant RCS for the INSAT series, on my initiative, ISRO entered into a contract with MBB-ERNO* of Germany during 1987–88 to supply INSAT-2A propellant tanks, and transfer the technology and tooling to manufacture these propellant tanks for the ISRO satellite programmes. As per

---

* Messerschmitt-Bölkow-Blohm (MBB), later taken over by DASA (Deutsche Aerospace AG) and currently part of the Airbus Group.

the contract, processes were established for the design and development of passive propellant management systems not only for the spacecraft but also for PS4, (the PSLV fourth stage solid motor). Existing facilities were augmented and new ones established to meet the production requirements of spacecraft tanks and PS4 propellant acquisition systems. We associated the expertise and facilities available in Indian industries, including BHEL and KCP, for the machining of thin pressure vessel shells, a large Coordinate Measuring Machine (CMM) for profile inspection, NDT facilities such as ultrasonic, fluorescent, eddy current and holographic inspections and a large EBW machine among other items.

For IRS-1D and subsequent monopropellant RCS for satellites, we used the metallic capillary acquisition system. In this context, C.S. Prasad of the SPSD, who coordinated the procurement of INSAT-2A propellant tanks and successfully executed the technology transfer programme from the MBB, deserves special mention. This technology transfer was useful not only for the manufacture of propellant tanks with a passive propellant management system for the satellites but also for the upper stages of the ISRO launch vehicle, including the PS4 propellant tank for PSLV. The successful on-orbit operation of monopropellant systems in IRS-1A, 1B and the Stretched Rohini Satellite Systems (SROSS) in the early 1990s, with 1N thrusters and some imported components, encouraged us to develop the 11N thruster and flow control valve for IRS-1C and 1D, including the indigenization of the monopropellant latch valve, thermocouples and system filters. Though several monopropellant RCS were successfully used in different satellites, its remarkable performance in the IRS-1D flight was noteworthy and memorable. It is briefly presented below.

The propulsion system for IRS-1D was realized with a totally indigenous system (except bed heater and transition joint) with four 30-litre surface tension tanks, sixteen 1N thrusters for on-orbit maintenance, In Plane Correction (IPC) and Out of Plane correction (OPC) in orbit, and one 11N thruster for correction in the orbit due to launch vehicle injection error. After all the preparations for a successful launch, the PSLV-C1/IRS-1D took off from SHAR on 27 September 1997, in the august presence of Prime Minister I.K. Gujral, Dr K. Kasturirangan who was the chairman of ISRO then, ISRO centre directors and team members.

The orbit realized was an elliptical one with the perigree at 301 km and the apogee at 821 km instead of the designed circular orbit with a radius of 821 km due to some deviation in the performance of the launch vehicle PSLV-C1. The onus was on the IRS-1D propulsion system to provide an additional velocity increment of about 140 metres per second to raise the perigee to 821 km. The mission strategy was worked out with four 1N thrusters and one 11N thruster, which demanded about 68 kg propellant for these manoeuvres, of which the 11N thruster used about 48 kg. The mission strategy was based on the qualification levels of the thrusters, thruster decay and roughness. My propulsion team made quite a few presentations to the satellite mission group and other experts. Based on detailed discussions and reviews, the team evolved an action plan to achieve the mission through the burning of different thrusters serially, tracking the orbit closely and making revisions in the plan as required from the response of the satellite.

Every second of the propulsion parameters was monitored. The first burn for the duration of 300 seconds was planned on 30 September 1997, with one 11N and four 1N thrusters,

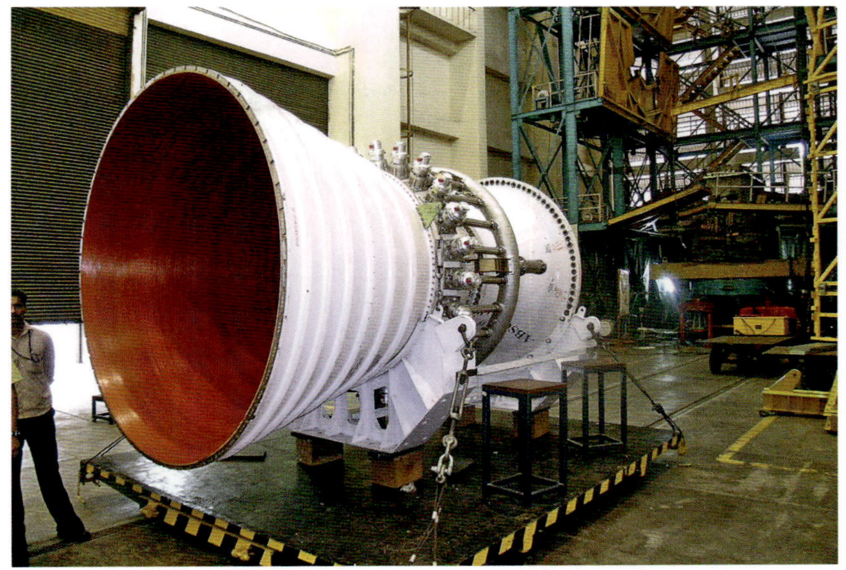

35. SITVC for PSLV first stage with twenty-four single pintle electromechanical injection valves, equally spaced around nozzle

36. Indian Remote Sensing (IRS) spacecraft

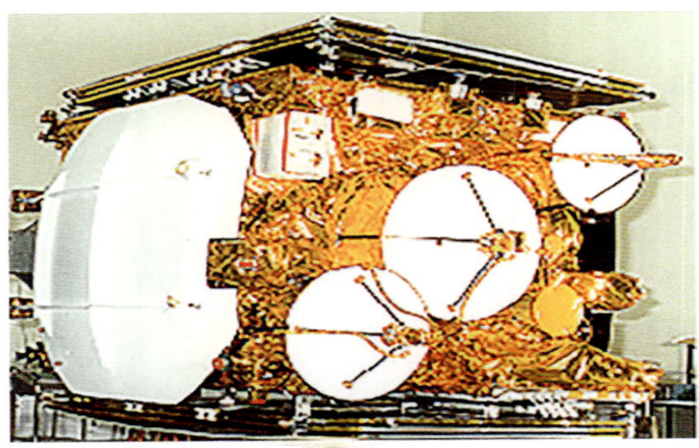

37. INSAT spacecraft

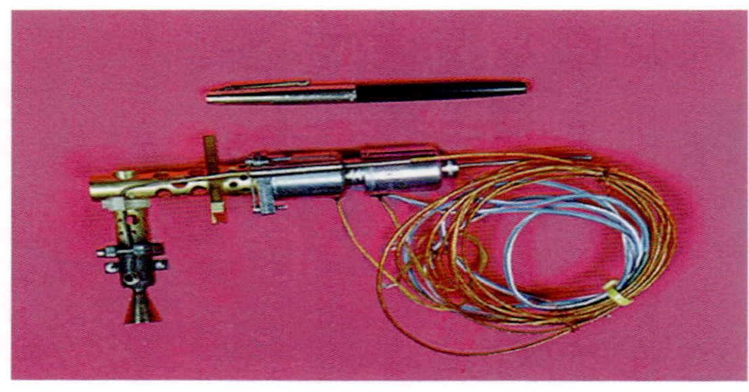

38. 1N mono-propellant thruster for IRS

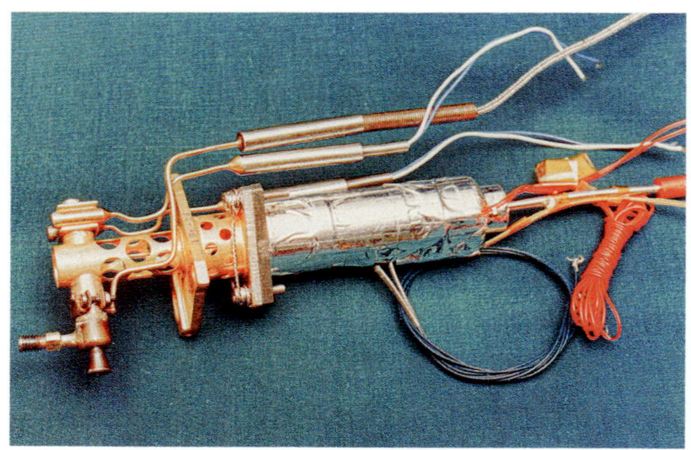

39. 11N mono-propellant thruster for IRS

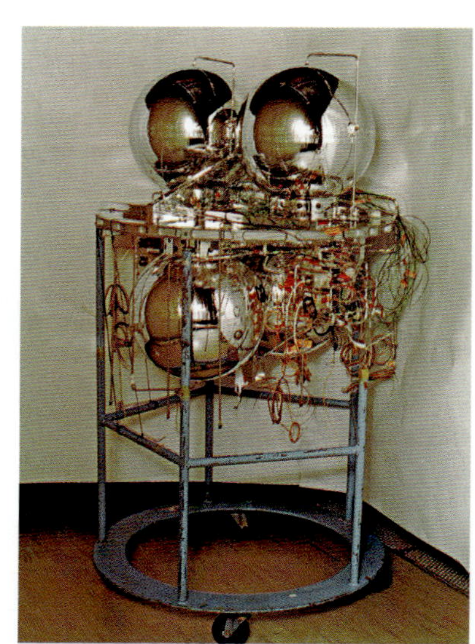

40. IRS propulsion system

and the next burn was scheduled for midnight. Every day, there used to be several burns involving very meticulous verification of the predicted thrust and Isp with what were realized. The series of manoeuvres for more than fifteen days, with the mission and the propulsion team virtually positioned at the ISRO Telemetry Tracking and Command Network (ISTRAC) campus in Bengaluru around the clock, raised the perigree of the satellite to 737 km, a satisfactory operational orbit to meet all mission requirements.

During the entire process, the 11N thrusters had a burn of 20,950 seconds, with a hydrazine throughput of 48 kg and only about 12 kg of propellant remaining in the system for the rest of the life of the spacecraft. The chairman of ISRO, the directors of the VSSC, LPSC and ISAC, and several others visited LPSC, Bengaluru, praised the IRS-1D RCS performance and applauded the team. After five years of the IRS-1D being in operation, 11N thrusters were again fired ten times for an additional 5,227 seconds in order to correct the inclination. It is heartening to note that IRS-1D completed twelve years of orbit life with the propulsion system delivering impulse till the last moment. Thus, the IRS-1D mission was saved, in spite of the underperformance of the launch vehicle, by the RCS that the LPSC had developed. The success was celebrated in a fitting manner with the participation of Dr U.R. Rao, our former chairman, and the directors and other ISRO personnel.

Further versions of the 11N thrusters were qualified and flown with limited pulse mode and total pulse mode capabilities for all duty cycles for different IRS spacecrafts. Propellant tank sizes also varied from 7.5 litres in Youth-sat to 390 litres for bigger spacecrafts. Valve designs underwent modifications with Teflon seats. By 2012, around twenty monopropellant RCS systems in seven different configurations were flown with a cumulative on-orbit life of more than 120 years with five spacecrafts that completed ten years in orbit. Recently, in the Technology Experiment Satellite (TES) spacecraft, the 100th manoeuvre was completed with 11N thrusters after fulfilling ten years of on-orbit life.

Bipropellant Propulsion System for Satellites

The major developments towards creating a bipropellant satellite propulsion system (Photo No. 41) were (a) the Liquid Apogee Motor (LAM) with a thrust of 440N to raise the orbit from geo transfer orbit (GTO) to geosynchronous orbit, and (b) the Attitude and Orbit Control System (AOCS) with 22N and 10N thrusters to provide the reaction control capability for active body-stabilized attitude control and orbit positioning of the spacecraft. A typical geosynchronous spacecraft has one 440N LAM and eight each of 22N and 10N thrusters. The development of LAM and AOCS started in the 1980s with initial sea-level development, followed by high-altitude development tests and qualification tests. The flight system was supplied for INSAT-2A, which was launched in July 1992. It included the development of associated flow control valves, fill/drain valves, pressure regulator, high-pressure gas tanks, fuel tanks and sensors. These aspects were complex and challenging tasks.

The Liquid Apogee Motor (LAM)

The LAM (Photo No. 42) developed 440N thrust using MON-3 and MMH propellants with a mixture ratio of 160. It had a coaxial swirl injector with oxidizer and fuel flowing through the outer and inner flow passages, respectively.

The radiation-cooled thrust chamber was made up of silicide-coated columbium alloy. Silicide coating activities were done initially by making use of the Defence Research and Development Laboratory (DRDL) facilities at Hyderabad as the LPSC did not possess the vacuum furnace that could be operated up to a soaking temperature of 1,350°C. The successful development of silicide coating on columbium alloy was a very arduous task. C.G. Balan and his team deserve special mention for this significant achievement.

Mark-I LAM was designed initially with an injection pressure of 15.5 bar. The spacecraft system pressure drop evaluation arrived at an injection pressure of 14.5 bar. The extra pressure drop was adjusted by increasing the radial gap in the swirl flow passages, and the engine was qualified and used in INSAT-2A, 2B and 2C. Owing to this adjustment, the specific impulse (Isp) came down to 307 seconds against the specification of 310 seconds (minimum). The design was modified to meet the revised injection pressure of 14.5 bar. With the revised design, one Mark-II engine (D3) was qualified in the High-Altitude Test (HAT) for a total burn time of 34,000 seconds with more than forty restarts. This version started flying from INSAT-2D and continues to do so even now (twenty-two engines). Even though the Isp specification was 310 seconds (minimum), Mark-II engines delivered a better (higher) Isp. This feat was achieved by the selective screening of injector flow elements and the specification is now revised to 315 seconds (nominal). It was observed that this value of Isp was the highest reported among this class of engines worldwide.

Internationally, the Mark-II engines were acclaimed to be of robust design as they had demonstrated operation in very harsh and volatile environments (pressure and temperature). Two engines of this version qualified for a total burn duration of 24,500 and 16,400 seconds, respectively, after subjecting them to the qualification-level thermo-vacuum and vibration tests. I wish the LPSC all success in its efforts to increase the Isp to more than 320 seconds by raising the area ratio from the existing 160 to 250.

AOCS Thrusters

(Photo Nos 44 and 45)

22N Thruster

This thruster is identical to LAM in terms of the type of injector and thrust chamber configuration. The injector is made of columbium-103 alloy unlike in LAM, which has a titanium-aluminium alloy injector. The injector configuration is similar to LAM with the oxidizer flowing through the outer flow passage and the fuel through the inner flow passage. Based on the performance in HAT, modifications were made in the design which included (1) modifying the seal between the valve and injector (2) interchanging the oxidizer and fuel flow path, and (3) changing the material of injector from columbium alloy to titanium-aluminium alloy. With these modifications, Mark-I was realized, qualified and supplied for use in INSAT-2A, 2B and 2C.

In the INSAT-2C spacecraft, several 22N thrusters showed a degradation in performance during the mission operations. This setback was attributed to the increase in pressure drop of the flow control valve due to lower stroke lengths. Accordingly, these valves were modified with higher stroke length, having an improved seat design, and the thruster was requalified and supplied for INSAT-2D. Even though Mark-I performed satisfactorily to meet the mission requirements, improvements in the performance were made in the Mark-II

41. INSAT propulsion system with Liquid Apogee Motor (LAM) firing

42. Liquid Apogee Motor (LAM) of 440N thrust

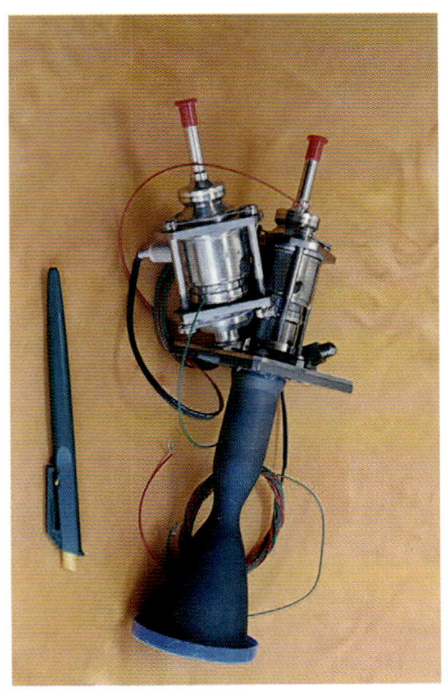

43. AOCS thruster of 10N thrust

44. AOCS thruster of 22N thrust

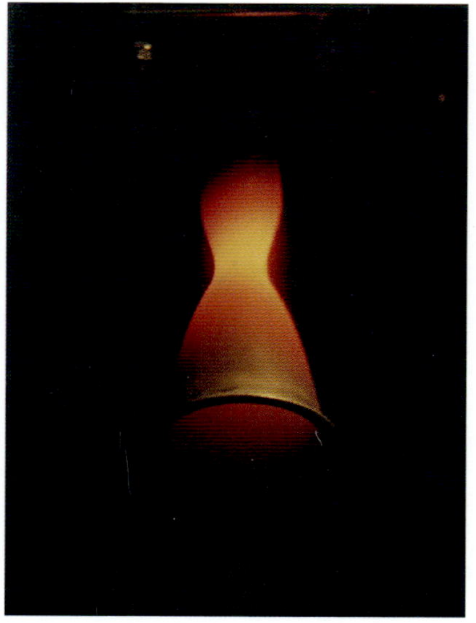

45. Long duration high altitude simulation test of 22N AOCS thruster in vacuum chamber

version in order to assure better pulse-to-pulse repeatability and hardware-to-hardware impulse repeatability by (1) increasing the pressure drop in the critical swirl flow path in the injector, (2) improved matching of dribble volumes, and (3) replacing the solenoid valve with the Latchable Series Redundant Valve (LSRV).

After qualification, the Mark-II version was used from INSAT-2E onwards. Even though the yield rate was better and the pulse mode performance also improved, this Mark-II version was further amended to the Mark-III version. This modification was done by interchanging the oxidizer and fuel flow path similar to the Mark-I version and optimizing the pressure drop to improve the atomization and increase the Isp to 290 seconds. The Mark-III version was inducted for flight from INSAT-4A onwards.

10N Thruster

This thruster (Photo No. 43) development was taken up based on the project requirement to develop a thruster with Minimum Impulse Bit (MIB) for better controllability and performance. It is similar to the 22N Mark-II version with the injector made of titanium-aluminium alloy and fuel flowing through the outer flow passage and oxidizer through the inner flow passage. The injector assembly was modified to interference fit, and the injector characterization trials were successfully completed. With the configuration given above, the first development hardware was tested for a cumulative firing duration of 7,045 seconds, including 40,050 pulses. Further, three more thrusters were qualified with a single seat valve, up to 252,000 pulses. This version of the thruster (eight of them) was supplied and used, in the east and west station repositioning and acquisition manoeuvres for the GSAT-1 spacecraft, in 2001 for the first time.

Normally, the AOCS thrusters are located at the corners of the satellite to take care of attitude hold, momentum dumping and station keeping. As the size of the satellite increases, the impulse requirement becomes less. So the 10N thruster was designed, developed and used in the east and west side of the satellite for attitude hold and momentum dumping, which are operated in pulse mode. Subsequently, one more thruster was qualified for 22,500 seconds, including 271,000 pulses. This version (eight of them) was used in fourteen spacecrafts – from INSAT-3C to GSAT-12. Further improvements in performance were planned.

The INSAT-2A propulsion system was configured based on the experience from INSAT-1A to INSAT-1D, which were built by Ford Aerospace with a single pressure regulator, a check valve, indigenous LAM, pyro valves, propellant tanks, bipropellant latch valves, a few AOCS thrusters and an imported pressurant tank. Subsequent to the INSAT-2A experience, with the modification of the regulator seat design, the system was configured with an additional pressure regulator in series and also by extra check valves and filters on the gas module side. The propulsion system underwent further changes over the years to overcome the anomalies and to incorporate redundancies, but we never had a LAM malfunction that led to mission failure.

As of 2012, more than twenty subsystems in different configurations have been flown in the INSAT/GSAT series of satellites, including one each in the foreign satellites – W2M and Avanti-Hylas. We qualified and supplied fill and drain/ vent valves and pressure transducers with a built-in amplifier to Hughes, USA. For Brazil, we designed, developed, produced and supplied ground support equipment to fill propellants in satellites. As mentioned before, we also conceived, realized and delivered propulsion subsystems for use in the satellites of the W2M and Avanti-Hylas projects. The propulsion system performance in these spacecrafts was highly appreciated, and there was never a constraint reported in the operation or life of the spacecrafts.

10
Cryogenic Propulsion and the Russian Connection

ISRO RECOGNIZED THAT THE development of cryogenic propulsion stages was essential for better future performance and the high payload capability of launch vehicles. After the excellent full-duration test of the Vikas engine at SEP on 9 December 1985, which demonstrated ISRO's ability to develop an earth-storable high-thrust liquid engine for at least its own launch vehicles, I desired to develop cryogenic propulsion technology in the Propulsion Group of VSSC.

Research and Development in Cryogenic Propulsion

To make a humble beginning in the Propulsion Group, I asked V. Gnanagandhi, who had knowledge of and exposure to the development of the Vikas liquid engine, and S.C. Ghosh and his group to begin the development of cryogenic technology with fresh, young engineers. At the time, the VSSC had initiated studies on cryogenic propulsion systems independently. In accordance with the directions of the ISRO chairman, some colleagues who were engaged in cryogenic system studies at the Propellant Group, led by E.V.S. Namboothiri and M.S. Bhat, joined my team to consolidate our R&D efforts. I established a cryogenic propulsion system development group, appointing E.V.S. Namboothiri as group director, V. Gnanagandhi as the group head and S.C. Ghosh as the divisional head.

Based on my recommendation, the then chairman of ISRO, Dr U.R. Rao, sanctioned a budget of Rs 18 crore towards the pre-project activity. After detailed studies, consultations with experts in the field and discussions within the group, a development philosophy was evolved for the indigenous cryogenic engine.

Initially, the team started working with a single coaxial element of gaseous hydrogen and gaseous oxygen, to study the ignition characteristics. Later, the engineers moved on to gaseous hydrogen with liquid oxygen. Dr K. Ramamurthi developed the ignition system of the cryogenic programme. The team progressed well under my guidance, and they designed a 1-tonne regeneratively cooled engine with nineteen injector elements.

Our first major initiative was the development of a regeneratively cooled thrust chamber using the electro-deposition technique developed by the Central Electro Chemical Research Institute (CECRI), a Council of Scientific and Industrial Research (CSIR) laboratory at Karaikudi. V. Gnanagandhi coordinated this development with CECRI. The regenerative channels for the initial 1-tonne cryogenic engine were manufactured, using a Computerized Numerical

Control (CNC) machine, under the guidance of Jones Rabindran. The liquid hydrogen we needed was not available in India. So the cryogenic development team proposed to produce liquid hydrogen by cooling gaseous hydrogen using liquid helium. Gnanagandhi discussed the matter with officials of the All India Institute of Medical Sciences (AIIMS), New Delhi, which had a liquid helium facility. Based on the discussions, a mini liquid hydrogen plant was established at ISRO in 1987. In this plant, the gaseous hydrogen produced through electrolysis was cooled to produce liquid hydrogen, which was supplied for the initial cryogenic engine tests.

Simultaneously, D. Sarveson, K. Alaguvelu, B. Chellathurai and J. Asir Packiaraj established a high-pressure test facility for the cryogenic engine in pressure-fed mode. Initially, in 1990, the 1-tonne cryogenic engine with water-cooled thrust chamber was tested successfully with gaseous hydrogen and liquid oxygen. Subsequently, in 1993, the regeneratively cooled 1-tonne engine was tested, unsuccessfully. Initially, there were failures due to non-ignition and explosions. After detailed failure analysis and based on the recommendations of experts, necessary modifications were made in the test procedures and helium was used to purge the hydrogen line.

After successful tests of a 1-tonne thrust water-cooled engine, I wanted to initiate developing a higher-thrust cryogenic engine which could find application as the upper stage for future launch vehicles of ISRO. Based on studies we conducted and consultations with the ISRO chairman, I decided to develop a high-thrust cryogenic engine with 12-tonne thrust. The cryogenic team designed the engine and initiated the realization programme. While this was in progress, I moved to the Department of Ocean Development as secretary to the Government of India in 1995. The cryogenic team continued the efforts, successfully testing the 1-tonne thrust regeneratively cooled engine using LOX and LH_2 in 1996. The team also realized a 12-tonne thrust chamber with electroforming route and a porous face plate in the injector. They successfully tested it in 1998 (Photo Nos 46 and 47).

For this significant achievement, I congratulate V. Gnanagandhi, S.C. Ghosh, Dr P. Balachandran, Dr V. Narayanan, S. Ambalavanan and Roy Pascal for their contribution in developing thrust chambers; N.K. Gupta, S. Arunan and M.S. Suresh for their work on the turbopump; N.R.V. Kartha on hardware realization in industries; and A. Amaraseharan for the flow control valves for the successful tests of C1 and C12 cryogenic engines.

Cryogenic System Project

During the last quarter of 1990, to execute the developments of a cryogenic stage in a project mode for GSLV, a cryogenic system project (CSP) was established at the LPSC with the project management board under my chairmanship and with S. Nambinarayanan as project director. Recognizing the complexity in the new development, I proposed to the chairman of ISRO a collaboration with 'Technology Transfer' – similar to ISRO-SEP Vikas agreement with France – with institutions abroad having the necessary expertise. During the 1970s, when ISRO had the collaborative programme with SEP on the Viking technology, there was a suggestion from the latter regarding yet another collaboration for the cryogenic third stage, HM-7, which was then under development for the Ariane programme. Since ISRO had just started a new initiative on liquid boosters with SEP, we did not pursue the idea then. Resource limitation was a major factor behind the decision.

However, in the 1990s, when ISRO planned a cryogenic stage development for GSLV, it approached SEP. Regrettably, the latter demanded a very high fee, and hence, ISRO did not pursue it. The cryogenic stage HM-7, under development for the Ariane programme, and the cryogenic stage RL-10 from Pratt Whitney of the US were options. ISRO felt it advantageous to approach Russian agencies since it had established good collaborations with the country in the satellite programmes. Accordingly, I proposed to the chairman of ISRO that we liaise with Glavkosmos (GK), the commercial wing of the Russian Space Agency, and associate with Salyut their Design Bureau for the cryogenic stage development and with KB KhimMash or the A.M. Isayer Chemical Engineering Design Bureau, Korolyov, Russia, for cryogenic-engine development.

As directed by the chairman, I negotiated the details of the collaboration with the Russian counterparts, in the company of representatives from ISRO and the GSLV project, and the agreement was signed in January 1991. The agreement had two components. The first was for the supply of two cryogenic stages for the initial flights of GSLV. The second concerned the transfer of manufacturing technology for the realization of the cryogenic stage in India as well as the associated training of Indian specialists in Russia. The mission management systems of the Cryogenic Stage (CS), including electronics for stage sequencing, thrust and mixture ratio control, steering engine gimbal control, tank pressure maintenance and so on were to be developed by ISRO. GK configured a cryogenic stage with a propellant loading of 12.5 tonnes of LOX and LH2 using a proto-engine RD-56 developed for the Russian lunar programme.

The cryogenic engine RD-56 was developed by the A.M. Isayer Chemical Engineering Design Bureau of Russia over a decade – from early 1960 to 1971 – for the N-1 rocket of USSR's lunar mission. This development was simultaneous with the American Apollo, which successfully landed on the moon in 1969. Unfortunately, the Russian lunar mission had four successive failures, and consequently it was abandoned. No further development of RD-56 has been reported. This cryogenic engine operated for 800 seconds at a chamber pressure of 55.9 bar with an expansion ratio of 198.7 and delivered a thrust of 69.6 kN with a specific impulse of 454 seconds, using LOX as oxidizer and LH2 as fuel with a mixture ratio of about 6.0.

After the ISRO–GK agreement, KB KhimMash modified the RD-56 engine, fine-tuned it to meet ISRO's CS requirements, renamed it KVD-1 and supplied it to ISRO. This KVD-1 engine assembly had a thrust chamber, pyrogen igniters, booster and main turbopump systems, active thrust and mixture ratio control systems, control components, pyro valves and two steering engines for vehicle control. The stage had aluminium alloy propellant tanks with foam insulation, structures, fluid circuits, components and instrumentation and control systems. While the engine was exclusively developed by Russia, the stage was a joint concern of Russia and India.

Cryogenic Stage: Procurement & Technology Transfer

As per the ISRO–GK agreement, I deputed a team – with V. Gnanagandhi as the leader – consisting of personnel from ISRO and Indian industries that were potential participants in the project. The team was deputed to the Russian facilities in a phased manner from January 1993 for training and realizing cryogenic technology acquisition. The other members of the team were Dr V. Narayanan, M.S. Suresh, G. Narayanan, Roy Pascal, P. Marudachalam, Sathish Chandra Kumar, N.G. Unni, A. Amaraseharan, M. Yagnanarayana, Dr K.V. Nagarajan, M. Sahabudin, M.S. Bhat, M. Xavier, D.P. Sudhakar, S.M. Rajaratnam, N.R.V. Kartha, Madan Lal and Jones Rabindran. This team also had two translators – S.D. Barman and B. Pandiyan.

Though the CS agreement was with GK, the actual development of CS took place in different facilities/organizations in Russia – the stage systems in the Salyut Design Bureau of Moscow, the engine systems in KhimMash (Moscow Region), engine systems realization in UKWP (UST-Katav, Wagon-Building Plant) of UST-Katav (Chelya Binsk Region), engine and stage testing in NIICHIMMASH (Scientific Research Institute for Chemical Machine Building) of Sergiev Posad (Moscow Region), and some stage systems in NIIMMASH (Scientific Research Institute of Machine Building) of Niznaya Salda (Sverdlovsk Region). The team members from ISRO were thus deputed to Salyut Design Bureau, KB KhimMash, UstKatav and NIICMMASH.

The team members had the opportunity to go through the drawings, the documents related to realization, and the assembly and acceptance test procedure. They got valuable exposure in cryogenic propulsion functioning, various subsystems and their functional requirements, pneumo-hydraulic fluid systems, engine control and aspects related to hardware realization, assembly, testing and qualification, and so on. The ISRO team also visited the test facilities at Foustava and NIICHIMMASH.

I recognized that hardware realization, based on technology transfer, was a very complex process. Large quantities of metallic and non-metallic materials had to be procured, which would have to be indigenized in a phased manner. Special purpose fabrication facilities needed to be established with equipment, including a five-axis CNC milling machine, vacuum brazing furnace for realizing combustion chambers, five-axis inducer milling machine, wire-cut Electrical Discharge Machine (EDM) for turbo-machinery realization, and sensors for the measurement of pressure, temperature, flow, vibration, heat flux and so on.

For these challenging tasks, I established a fabrication team under the leadership of D. Sasikumar, who took over as the deputy project director of CSP, to plan and interface with Indian industries, and for the realization of cryogenic system hardware in close coordination with S. Nambinarayanan. Sasikumar was a mechanical engineer with years of experience at the VSSC and in other industries. With the approval of the chairman of ISRO, and the director of the VSSC, he was transferred to LPSC to take up the challenging assignment. I recall that Sasikumar played a very significant role in realizing the SLV-3 hardware in Indian industries under the guidance of A.P.J. Abdul Kalam. He demonstrated exceptional capability in the indigenization of materials, interfacing with industries and meeting project schedules.

During this period, a number of senior engineers from my team – E.V.S. Namboothiri, D. Sasikumar, K. Kasi Viswanathan, S.C. Ghosh, N.K. Gupta, Mohammed Muslim, Lekshmi Narayanan, Dr K. Ramamoorthy, R. Krishnan Potti, K. Algavelu, D.R. Selvanayagam, A. Narayanan and Baby T. Mathew – were deputed for a short period to visit the Russian facilities and discuss matters with Russian experts. They were accompanied by representatives from Godrej & Boyce, Mumbai, MTAR Technologies, Hyderabad, Larsen and Tubro, Mumbai, Walchand Industries, Walchand Nagar, Mumbai, HAL, Bengaluru and KELTEC, Thiruvananthapuram. The team had the support of the embassy of India during the period of deputation.

In April 1993, a high-level ISRO team led by Dr U.R. Rao and senior officials – including the director of LPSC, the project directors of GSLV and CSP, and the directors of ISAC and LVPO – visited Russia and reviewed the progress of the activities. I went to the work centres at Salyut Design Bureau, KB KhimMash and UST-Katav along with Gnanagandhi and GSLV representatives and held discussions with Director Leontiev and Deputy Director Selesniev of KB KhimMash and Kirlivchev, director, UST-Katav, to subsequently take major decisions on the procurement of material for engines and specific machines such as the five-axis machine and the spline milling machine required for the development in India. I also held talks with Technomash regarding the establishing of a brazing furnace in India. In addition, there was a detailed review of technology acquisition by individual engineers on different subsystems, including the thrust chamber, turbopump, valves, tanks, fill and drain system, umbilical system and control system, among others. I subsequently visited Glavkosmos in connection with a techno-commercial meeting.

The ISRO team at Moscow stayed in rented apartments. They cooked their own food and had Indian meals whenever they felt like it. I visited our team in their apartment, had a brief review meeting about their activities at the work centres and enjoyed the Indian food. Our team members availed of public transport when the industry could not pick them up for work. At times, they used private vehicles which operated as taxis. Gnanagandhi told me that private vehicles going in the same direction would drop them off for a reasonable charge. Once, two of my colleagues stopped a private vehicle and asked for a lift to my hotel along with me. The driver obliged and picked up all three of us. The charge was reasonable.

Along with Gnanagandhi, I called on Ronen Sen, the ambassador, and briefed him about our plans and activities. He was very responsive and offered all possible help. He cautioned us from moving around alone, particularly at night.

While we were talking, some firing was reported near the hotel where I was staying. There was some unrest and the army was called into action. The white building where the parliament of Russia met and several high-level government offices were housed was attacked by rebels. The hotel I was staying at was at a distance of less than 500 metres from the white building. There was firing in the night and one person was shot down by the armed forces. Early the next day, the embassy arranged for my travel to the airport in one of its official vehicles and took me through the back door of the hotel to ensure my safety.

I observed that V. Gnanagandhi had organized the acquisition of the cryogenic technology very well and that the team members worked together as a family. Our employees worked more than ten hours a day on an average to bring back maximum technology from Russia. During this period, the Russian economy was down and America pressurized Russia to not give cryogenic technology to India.

ISRO–GK Agreement Rescinded, Renegotiated and Modified

While the technology acquisition programme was in progress, a political decision was taken by Russia to rescind the agreement signed in 1991 and not transfer the cryogenic technology to ISRO. Along with my colleagues, I renegotiated the agreement of 1991 and got it modified in December 1993, deleting the technology transfer component but retaining the supply component with amendments. According to the new agreement, GK would supply to ISRO, in addition to the two flight CSs, one mock-up model for propellant filling, one technological mock-up for handling studies, and two more flight CSs as compensation. Moreover, Russia would give three more flight CSs to us for a mutually agreed additional prize. Subsequently, it was decided that the last two stages would be upgraded to 15 tonnes of propellant loading and thrust of 95 kN to enhance the payload capability of GSLV. In total, seven flight-worthy stages had to be supplied to ISRO.

I assessed in detail the status of the technology acquisition and consulted every team member deputed in Russia to recommend to the chairman of ISRO the future strategy for cryogenic developments since Russia had denied the technology transfer. Based on assessment and inputs from the ISRO team in Russia, I recommended that we continue with the cryogenic stage development in India with the knowledge so far gained. I requested the chairman to assign to the team members trained in Russia suitable responsibilities in the modified cryogenic development programmes so that ISRO could benefit from and gainfully utilize the expertise that had been acquired.

During the CS development in Russia, a large number of ISRO specialists got the opportunity to understand cryogenic systems and the interaction between propulsion and control systems. ISRO experts also took part in the development and flight acceptance hot tests of engines and the acceptance of flight CS. The technological mock-up and fluid mock-up were delivered to ISRO in April 1997 and October 1998 respectively. After completing the development of CS, the first flight CS was delivered to ISRO in September 1999 and the remaining six stages were supplied in a phased manner between July 2000 and September 2007.

A large number of GK specialists were involved in successfully realizing and commissioning the launch complex facilities at SDSC. The first flight CS was prepared in the technical complex established at SDSC using the equipment and pneumo-control panels supplied by GK and integrated with the rest of the GSLV-D1 vehicle systems. The first GSLV MK-II vehicle equipped with Russian-supplied CS was successfully launched on 18 April 2001. This great triumph was the result of cooperation between ISRO and Russia, and both the teams were proud of this accomplishment. Since then, ISRO has accomplished five more GSLV MK-II launches with CS in three successful missions.

Cryogenic Test Facilities

In 1992, I visited the Research Institute of Mechanical Engineering (RIME) of Salda, a Russian aerospace firm, discussed a development plan to establish cryogenic systems assembly and test facilities at Mahendragiri, and entered into an agreement for technical assistance with it in 1993. The facilities to test the Cryogenic Upper Stage (CUS) propulsion system were established at Mahendragiri under the guidance of Russian experts. RIME extended its cooperation in the design, engineering, fabrication and supply of cryogenic fluid systems, including super-insulated pipe lines, super-insulated valve units, and the disposal circuits meant for handling LH_2 and LOX. For the assembly and integration, testing and qualification of CUS, various facilities such as cold flow and hot test facilities were established at the ISRO Propulsion Complex (IPRC). An integrated liquid hydrogen plant (ILHP) producing 500 kg of LH_2 per day was also established and commissioned in August 1995.

Cryogenic Upper Stage (CUS) Project

I left my assignment at ISRO in April 1995. To continue with the development of the CUS development for GSLV, a CUS project was established at the LPSC in 1995. Gnanagandhi was appointed as the project director of CUS, with a project management board under the chairmanship of E.V.S. Namboothiri. I am informed that the CUS project established in Godrej & Boyce, MTAR Technologies and HAL, special facilities and infrastructure for the realization of engine and stage systems. The hardware for the development for GSLV was realized in Indian industries, and the CUS was successfully tested for the first time in early 2002 at Mahendragiri.

This development had overcome several technological hurdles. The major challenges were in realizing cryogenic engine hardware for the combustion chamber, turbopump and control valves. Since the density of hydrogen is very low, it is liquefied at a temperature of 20K (i.e.,–253^0 C) and stored in stage tanks. This liquid hydrogen (LH_2) cannot be injected directly into the combustion chamber. It has to be heated and converted to gas prior to the process. The combustion chamber had a copper inner wall with a high-strength stainless steel outer jacket. Helical rectangular channels with varying width and helical angle were milled outside the copper chamber. Programming the required algorithm with three rotary and two linear axes was a great challenge to manufacture. After the successful manufacture of the copper inner chamber, joining it by brazing to the outer stainless steel was a tougher task.

For brazing, a special rotary vacuum furnace was established at Godrej & Boyce, Mumbai. Brazing was carried out in an inert atmosphere at a pressure of 2 bar and temperature of 1020^0 C. A vacuum of 10^2torr* is created in the channel between the copper and stainless steel outer jackets. The assembly was rotated at 1–2 rpm. When the first brazing was completed, it was observed that the outer stainless steel jacket cracked like a mirror in all directions. This problem was analysed in depth and it was found that the failure was due to the very high stress during brazing, which arose out of the differential expansion between copper and stainless steel. The problem was overcome by providing the flexures to facilitate the differential expansion.

The realization of the injector was yet another huge obstacle due to multiple coaxial injectors arranged in rows, separated by four compartments without any intercommunication. Every single element had a LOX port with a tangential entry hole to enable LOX to swirl and come out in the form of a fan. Similarly, the hydrogen passed through a helical vane to impart tangential velocity and come out as a fan. All these elements were meticulously fabricated, water-calibrated, assembled with braze foils and brazed as a single

* Torr is usually used in vacuum systems having very low pressures. 1 torr is 9.81 milli bar.

unit. Further, the turbopump, which fed the propellant to the combustion chamber, ran at 40,000 rpm. It had a single turbine, two-stage hydrogen pump and a single LOX pump, with a separating seal unit between the pumps. The turbine vanes were closed ones. Hence the vanes were machined by the EDM process from the sides to get the proper profile. It was a great challenge to get a correct profile without a top opening. Since the turbopump assembly ran at 40,000 rpm, dynamic balancing had to be done, meticulously, in steps.

The whole cryogenic stage required more than sixty valves, including command valves, injection valves, check valves, relief valves, purge valves and metering valves for proper functioning of the subsystems. The valves had to function at a very low temperature without any internal leakage. Hence the selection of seals, springs, bellows and assembly with close tolerances was a challenge. The major development of the systems took place under the leadership of V. Gnanagandhi. Subsequent to his superannuation in January 2006, Mohammed Muslim led the project till the first launch in 2010.

After carrying out extensive development and qualification of the engine and stage, the first flight stage was realized and tested in the GSLV-D3 flight on 15 April 2010. The engines successfully ignited but could not develop the required thrust, and CUS could not perform due to the malfunctioning of the LH_2 booster turbopump. After detailed failure analysis, corrective actions were taken and the GSLV-D5 flight with CUS was a grand success. Subsequently, GSLV was made operational with CUS.

ISRO continued with the development of heavier cryogenic stages for its later missions. After the successful CUS used in GSLV, ISRO developed C25, a cryogenic stage with 28 tonnes of propellant, powered by CE20, a cryogenic engine developing a 20-tonne thrust for use in the GSLV-Mk-III vehicle. Unlike CUS, the C25 development was totally indigenous. The C25 project team was led by D. Sarveson, N.K. Gupta and C.H. Kunnikamaran during the initial phases. The final phase of the development of C25 stage was executed by Dr V. Narayanan as the project director. The team successfully completed the development and realized the first flight stage. The first development flight (D1) of the GSLV-Mk-III, using this stage, was a grand success in its maiden attempt in July 2017. Currently, GSLV-Mk-III is an operational vehicle, to be deployed for Chandrayaan-2, Chandrayaan-3, and Gaganyaan missions.

I recall our humble beginnings in the development of cryogenic propulsion systems during the second half of the 1980s and am delighted to learn today about the significant achievements and demonstrated capabilities in the area. The experimental studies initiated on cryogenic propulsion systems during 1990 in CSP, and the knowledge and experience gained through C1 and C12 sub-scale programmes, gave very valuable inputs for the development of the C25 cryogenic stage and CE20 cryogenic engine. We proudly recognize that the 'technology denied to India' has now been 'developed in India' by my erstwhile cryogenic team. They have mastered system engineering, design, manufacturing, quality assurance, safety, infrastructure and facilities for assembly, integration, testing and evaluation. I believe that the technologies developed and the human resources established will meet all the challenges of the present and future ISRO launch vehicles and satellite programmes. I congratulate the LPSC and wish them all success in their future endeavours.

11

Components: Control Systems and Liquid Stages

THE DEVELOPMENT OF COMPONENTS for the control systems and liquid propulsion system had been a major activity of my LCSD team since 1968. When the development of the liquid propulsion system was initiated with a focus on small thrust engines (thrusters) capable of producing a 5 to 10 kgf thrust, the programme was plagued by the non-availability of proper fluid control components such as propellant flow control valves and pressure regulators.

Launch Vehicle Control System and Components Division (LCSD)

The testing of liquid propellant thrusters needed remotely operated, fast-response solenoid valves to admit or cut off the flow of toxic and hazardous propellants, for the safety of personnel and the test facility. Pressure regulators were needed to maintain the gas pressure in the propellant tank, so as to keep the engine thrust constant while firing. In the absence of proper fluid control components, hot tests of the engines had to be conducted with manually operated needle valves and commercially procured pressure regulators in the fluid lines, despite the attendant possibility of hazards to operating personnel.

These factors, as well as the expected demand for many types of small thrust engines for SLV-3 control systems, stimulated parallel development activities related to fluid control components in the Propulsion Engineering Division (PSN) and in the Control, Guidance and Instrumentation Division (CGI) from 1968 which were led by M.K.G. Nair and Sridharan Dhas, respectively. K. Radhakrishnan, Baby T. Mathew, J. Watson, S. Palaniappan and H. Vankateswara. Iyer assisted Sridharan Dhas in the development of the SITVC system and fluid components at CGI. In the PSN, an exclusive team headed by M.K.G. Nair was formed and given the responsibility of designing and developing fluid control components for liquid propulsion systems. E.S. Joseph, T. Jeyakumar, C. Amarasekaran, P.V.O.K. Nair and C. David – the last being a smart technician – were part of this team in its formative phase.

A temporary hangar near the Propellant Engineering Division (PED, the present Polymer Complex at the VSSC) had been partitioned into four labs that catered to various propulsion-related R&D activities. The components lab, measuring about 20 ft x12 ft, was one of them. An assembly and test bench, a DC power supply, a nitrogen cylinder, a few racks with flexible hoses, tools, adaptors and some stainless steel rods were all that the laboratory had at the time. M.K.G. Nair was seated in the narrow corridor in front of the lab.

Other engineers did not have proper offices. They used an armless chair and a stool in front of M.K.G. Nair's table in a 'musical chair' fashion. This team originally developed solenoid valves and dome-loaded pressure regulators to cater to the demand arising from 5 kgf and 10 kgf bipropellant thruster development.

In 1971, two more engineers D. Subramanian and R.C. Mahajan, and a few technicians – Sethuraghavan (turner), Hameed Rawther (fitter) and Syamalambika Kumari (coil winder) – joined the team. Since the space in the components lab was insufficient to accommodate the additional manpower as well as the machines, the components team was given a new larger hangar area near the SSTC main building. The first machines of the components workshop – an HMT lathe and a drilling machine – were soon commissioned to minimize dependency on city workshops for the fabrication of component prototypes.

While E.S. Joseph worked on the development of solenoid valves and dome-loaded pressure regulators, Jayakumar took charge of the fabrication and also worked on facility development. C. Amarasekaran designed relief valves and pressure switches, apart from working in the assembly and testing of components. He also built component test rigs with the assistance of Hameed Rawther, D. Subramanian and P.V.O.K. Nair, and procured the electronic equipment needed for the components instrumentation lab. R.C. Mahajan was in charge of solenoid coil winding and Syamalambika Kumari assisted him.

The solenoid valves, relief valves and dome-loaded pressure regulators developed by the components team were successfully used in the 600 kg liquid propellant engine development tests and the stage flight test in 1973.

T.R. Sundaramurthy, a new mechanical engineer, joined the components team in this period to assist Joseph in the design of solenoid valves for the second stage bipropellant reaction control system (RCS) and the third stage monopropellant RCS for SLV-3. Sundaramurthy was an excellent design engineer, and he introduced a standard method for the design of the solenoid valve electromagnetic circuit. Thanks to his active and crucial contribution, the design of solenoid valves for the RCS of the second and third stages of SLV-3 made good progress.

In 1974, M.K.G. Nair, E.S. Joseph, Jayakumar and Baby T. Mathew had to leave for SEP, France, on a long-term mission to acquire the Vikas engine technology. A.P.J. Abdul Kalam, the project director of SLV-3 then, was worried about the 'brain drain' in the components area and its likely impact on his project. He brought his concern to the notice of Dr Brahm Prakash, the VSSC director, and me. To address this issue and consolidate the control systems and components activities under one roof, N. Sridharan Dhas and his team were transferred from CGI to my division. Meanwhile, in 1975, Sundaramurthy resigned from ISRO in order to take up a job at the Kalpakkam Fast Breeder Reactor Project, and hence, C. Amarasekaran was given the responsibility of continuing the development of solenoid valves for SLV-3 control systems in addition to his other duties.

N. Sridharan Dhas was a dynamic leader and taskmaster, and he soon elevated the components activity to a new level by infusing more manpower and building crucial facilities. Initially, ISRO and the VSSC were under a registered society. During 1976–77, when both organizations came under the Government of India, the components activity was shifted back to the hangar near PED, this time occupying the whole hangar area. The space was partitioned to house a drafting room, fabrication machinery, assembly and test lab, and seating for the engineers.

By 1977, the PSN was reorganized into the propulsion group and Sridharan Dhas became head of the Propulsion Control Systems (PCS). C. Sundaram, a new engineer, and draughtsman T.V. Gopalakrishnan joined the components team. More engineers – M. Sundararaman, M. Yagnanarayana, Thomas Jayaseelan, N. Jayapandian, C. Johnson and A.S. Sivaramachandran – were redeployed to the components team in this period. Sundararaman developed the process for the manufacture of Teflon bladders for the propellant tanks of the SLV-3 second stage RCS. He also designed and developed a reliable and leak-tight check valve for the system. M. Yagnanarayana designed the spring-loaded pressure regulators for all the control systems of SLV-3. A.S. Sivaramachandran worked on pneumatically operated valves. He developed a pneumatically actuated start valve for the SLV-3 SITVC system, as a replacement for a pyro valve, which caused shock-induced dynamic problems when fired open.

The period between 1976 and 1979 was action-packed for the components team. As the components production for the ground qualification tests of SLV-3 control systems picked up, Sridharan Dhas expanded the components infrastructural facilities in a big way, after shifting it to a new, larger area opposite the present multi-storey building. Assisted by Thomas Jayaseelan, N. Jayapandian and C. Johnson, he changed the face of the components workshop by adding several new machines, including precision Schaublin lathes, jig boring machine, Studer grinding machine and milling machines. Skilled turners, machinists and assembly fitters such as Paradesi Asari, A. Thiruppathi, Kulandaivelu and S. Padmanaban were recruited/redeployed to man the facilities.

V. Ravi, a new engineer, set up an inspection laboratory for components parts. A high-pressure gas storage facility, pneumatic test facility, water flow test facility, solenoid coil winding lab and an instrumentation laboratory were built with the active support of J. Watson, K. Radhakrishnan, C. Amarasekaran, H.V. Iyer and P.V.O.K. Nair. Baby T. Mathew played a major role in establishing a Class 10,000 cleanroom for the assembly of components, in the basement area of the VSSC main building. H.V. Iyer was given the responsibility for components assembly and K. Radhakrishnan managed the instrumentation, testing and evaluation activities.

The qualification of fluid control components for the SITVC system and RCS of SLV-3 was completed successfully and fifty-six flight components were delivered during 1978–79. However, the maiden flight of SLV-3 in 1979 failed due to the suspected malfunctioning of a 250 kgf thruster oxidizer flow control solenoid valve, which formed part of the second stage RCS. An alternative cause of failure and the suspected failure of the switching relay were proposed. The failure analysis taught the components team valuable lessons.

The team converted this setback into an opportunity to revisit every aspect of the solenoid valve design, identify weak links and decide on corrective actions. Consequently, the solenoid valves were totally redesigned to improve their mechanical and electrical reliability and were requalified successfully in record time, thanks to the dedication of the team members who worked day and night, with the workshop running in three shifts. The performance of all the components was satisfactory in the next successful flight of SLV-3 in 1980 and also in the continuation flights. Since then, fluid components have never been the cause of failure for any ISRO mission so far.

During 1979–80, senior components engineers returned to India after the successful completion of the Vikas engine technology acquisition programme. They brought elaborate documentation on the Ariane-4 second stage fluid components and modules. The year 1980 saw the metamorphosis of the PSN group into two major units – the Auxiliary Propulsion Systems Unit (APSU) and the Liquid Propulsion Projects (LPP) – and I headed both. The Launch Vehicle Control Systems and Components Division (LCSD) was formed in APSU, and Sridharan Dhas was designated its head. E.S. Joseph became the head of the components section under LCSD. Baby T. Mathew coordinated fabrication, assembly, testing and delivery of components for the continuation flights of SLV-3. M.C. Indira was redeployed to the components team to work on development valves for the SITVC system and spacecraft propulsion.

Consequent to the finalization of the ASLV configuration during 1982–83, C. Amarasekaran and M. Yagnanarayana developed many components using new technology and materials to meet the ASLV requirements of SITVC and RCS. While Amarasekaran developed and introduced the latch valve technology and magnetically efficient solenoid magnetic materials, Yagnanarayana developed a cold gas RCS with 50 kgf thrusters for the AS0 stage. He was also given the overall responsibility for the realization and delivery of components for ASLV flights. As many as 110 components were required for each ASLV flight, for use in the control systems of AS0, AS1, AS2 and AS3 stages. Although the first two ASLV flights, in 1987 and 1988, were failures, control

46. First regeneratively cooled thrust chamber of 12 tonne thrust cryogenic engine with Liquid Oxygen-Liquid Hydrogen propellants, assembled on the test bed at Mahendragri

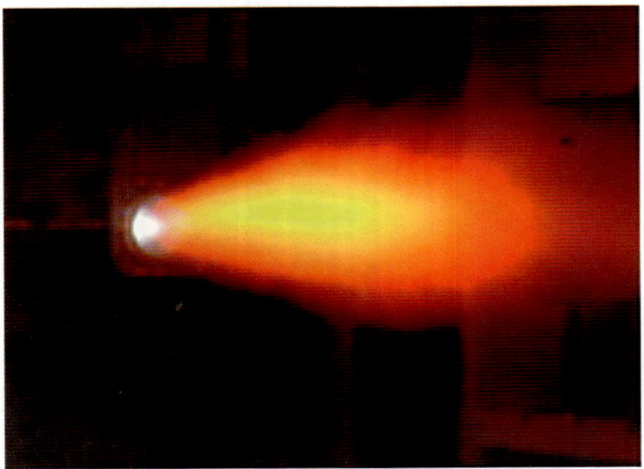

47. First successful static test of 50 seconds duration of 12-tonne thrust cryogenic engine in 1998 at Mahendragiri

48. Command System Module (CSM)

49. Pressurization Module-Group I for PSLV first stage reaction control system RCS) and fourth stage

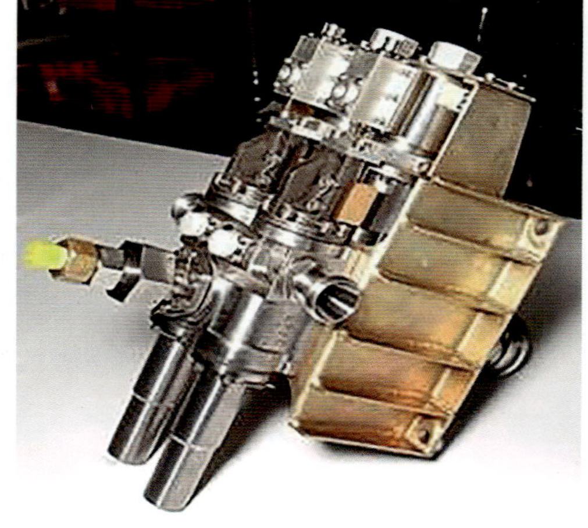

50. Hot Gas Roll Control Module (HGRCM)

systems and components worked satisfactorily in the flights as well as subsequent successful missions.

In 1985, APSU and LPP were merged to form the Liquid Propulsion Systems Unit (LPSU). In the same year, the configuration of PSLV was finalized with liquid propulsion systems for the second and fourth stages. The configuration for INSAT-2A, the first indigenous communication satellite, was also defined around this time, with a unified liquid propulsion system. PSLV required as many as 221 components for its liquid stages and the first stage SITVC system; the INSAT-2A propulsion system needed forty.

The subsequent years saw the highest workload for components engineers as they had to design and develop a large number of hi-tech components for the liquid propulsion systems of PSLV and INSAT-2A simultaneously. Baby T. Mathew developed the Pressurization System Modules (PSMs), and E.S. Joseph worked on the command system module (CSM) for PS2 (Photo No. 48). Amarasekaran developed the Main Engine Valves (MEVs) and PSMs for PS4 (Photo No. 49), apart from the solenoid valves and latch valves for the PS1 SITVC system. Yagnanarayana developed the Hot Gas Roll Control System Module (HRCSM) for PS2 (Photo No. 50), in addition to the PS1 SITVC system isolation valve module and PS4 pressure regulators. M.C. Indira developed the torque motor actuated injection valve for the PS1 SITVC system, which used twenty-four such valves. D. Venkittaraman aided her in this effort.

The design of fluid components and modules for PSLV liquid stages was a difficult task in view of the modularization requirements, high operating pressures and stringent specifications for leak tightness with helium gas. The PS1 SITVC injection valve needed a high-performance pancake torque motor and precision ball screw for efficient conversion of the motor rotation into the linear movement of the valve pintle. It took several design iterations to finalize the configuration of PSLV components and modules. But a bigger challenge had to be faced in the design of INSAT-2A components, owing to the need for a twelve-year propellant compatibility, near-zero leakage, very low power consumption, millions of life cycles, redundancy and miniaturization.

While Amarasekaran designed and developed the propellant flow control solenoid valves for the LAM engine and AOCS thrusters, Yagnanarayana designed and built the helium gas pressure regulator and check valves for INSAT-2A. Using his SEP experience, E.S. Joseph introduced standardization and a systematic approach in the procedures for design, drawing, assembly, testing and technical documentation in the components section. This step was very much needed at a time when the number of projects in ISRO and types of fluid components to be realized in LCSD were increasing every year.

Considering the huge amount of work involved in the development and realization of components for PSLV and INSAT-2A, additional facilities and manpower were sanctioned by the department for design, fabrication, assembly and test activities of the components section. Young engineers, including A. Manimaran, M. Radhakrishnan, G. Sundaravadivelu, Poly P. John, S. Vinodkumar, S. Jagannathan and M. Sreekumar, joined the components team and supported the components design and development activities for PSLV and INSAT-2A with dedication.

During the decade beginning in 1984, huge LCSD facilities and office buildings came up in Valiamala, thanks to the untiring efforts of N. Sridharan Dhas and his team. The components design team, made up of ten engineers, was the first liquid propulsion community to land in Valiamala, which was a virgin forest of breathtaking beauty at that time. The place looked more like a picnic spot than a space centre complex. Telephone facilities had not been installed yet. The quiet and lovely ambience was quite conducive to uninterrupted thought and concentration. Many pioneering design concepts used in the fluid components and modules of the PSLV liquid stages (PS2 and PS4) and the INSAT-2A propulsion system evolved during that period. In 1987, the LPSU was upgraded to the status of a centre and renamed the Liquid Propulsion Systems Centre (LPSC), with myself as its director. At the LPSC, components activity came under a new group called the Vehicle Control Systems and Components Group (VCCG). Sridharan Dhas became its group director, and E.S. Joseph was designated head of the Fluid Components Division (FCD).

The development and qualification of components and modules for PSLV liquid propulsion systems witnessed several technical problems related to regulator instability, excessive dynamic loads, O-ring blowout, seat leakage and so on, mainly due to high operating pressure (330 bar). These issues were resolved without affecting the components delivery schedule. However, INSAT-2A was a different story. A number of performance anomalies and failures cropped up in the AOCS thruster flow control valves and pressure regulators, not only during ground qualification but also in the orbit control operations in space. During the INSAT-2A mission in 1992, at the time of the first orbit-raising manoeuvre, the helium gas pressure regulator failed in open condition, on account of shock generated during the firing of the upstream pyro valve, which was meant to pressurize the propellant tanks. As a result, the tank pressure started increasing uncontrollably. So did the blood pressure of the mission controllers and observers.

Fear of an impending propellant tank explosion and the failure of the first indigenous communication satellite mission was writ large on all faces. As the tank pressure was crossing its maximum design limit, mission controllers intuitively fired the LAM as a desperate measure, much before the planned time, so as to allow the propellant tank pressure to reduce to a safe level. The INSAT-2A mission survived. It took multiple failure simulation tests, spread over many days and nights for months, to identify the shock and debris generation during pyro operation as the root causes of the on-orbit anomaly. The pressure regulator design was improved with many safeguards to withstand the shock and debris generated due to upstream pyro valve firing. In addition, the satellite propulsion system was also modified by introducing series redundancy for regulators. The regulator performance was satisfactory in subsequent INSAT missions.

Amarasekaran and Yagnanarayana were part of the team that went to Russia in 1993 to acquire the cryogenic engine technology. Unexpectedly, the technology acquisition could not be completed mainly due to US pressure. However, using the knowledge gained in Russia and their two decades of design experience at ISRO, components engineers developed and qualified the fluid components and modules for the cryogenic engine and stage successfully. However, it must be admitted that the exposure to Russian technologies, interaction with Russian specialists and visits to Russian industries during the Russian technology acquisition programme greatly helped in shortening the development time needed for realizing cryogenic components.

In 1994, N. Jayapandian, who was managing the in-house as well as external fabrication of components, died after battling cancer for two years. C. Johnson took over as the manager and new engineers – A. Alex and A. Manickavasagam – joined him. With the commencement of PSLV flights from 1993 and the formation of GSLV and cryogenic upper stage (CUS) projects, the demand for components and modules increased. The GSLV required 367 components for its liquid propulsion stages. In-house facilities were under strain, in spite of additional machines and equipment. With a view to augment the production rate of components by tapping the surplus capacity of external industries to the maximum extent possible, the Components Production Project (CPP) was formed in 1993. Baby T. Mathew was designated the project manager of CPP, in addition to his design responsibilities related to PSLV and GSLV liquid propulsion components and modules. Assisted by C. Sundaram and R. Rajasekaran, his team realized the components in large numbers through external fabrication and stocked sufficient quantities of them, so as to ensure the smooth and uninterrupted assembly of components in the in-house facility. H.V. Iyer and M. Sreekumar looked after component assembly and testing. P.V.O.K. Nair managed the environment test facilities and instrumentation. The GSLV-L40 stage components and modules developed by Baby T. Mathew and A. Manimaran performed well in the stage hot test in 1996.

In 1997, a month after the launch of INSAT-2C, a few 22N AOCS thrusters malfunctioned without producing the designed thrust. On-orbit measurements indicated that the oxidizer flow control solenoid valves in these thrusters either did not open or opened only partially. Extensive simulation tests and accelerated propellant compatibility tests on the ground traced the anomaly to the 'swelling' of the Teflon seal used in the solenoid valve seat, owing to the permeation of the oxidizer N_2O_4. The valve configuration went through several cycles of modification and requalification, in view of the persistent problems, after every redesign attempt. Many issues related to propellant compatibility, vibration levels, flexure fatigue failure, leakage and coil insulation degradation were resolved through sleepless nights of re-fabricating and retesting, until a robust valve configuration was finalized and qualified.

Similarly, during 1998–99, a miniature solenoid valve developed for the 10N thrusters of the INSAT-3 series spacecraft exhibited serious dynamic problems during qualification-level vibration tests, consequent upon the coupling of natural frequencies of valve and thruster assembly. Even after the solenoid armature and its flexure support configurations were changed several times, a solution was not in sight. Only after the armature was drastically modified to reduce its mass and increase its natural frequency, did the vibration-related problems vanish and the qualification tests become successful and complete. Since then, the performance of this valve has continued to be satisfactory in a number of spacecrafts, including Chandrayaan-1 and Mangalyaan.

During 1998–99, a new entity named Spacecraft Propulsion, Control Systems and Components (SPCC) was formed within the LPSC, thereby merging the activities of spacecraft propulsion in Bengaluru and the control systems/ components division in Thiruvananthapuram. N. Sridharan

Dhas was designated the associate director of SPCC, and E.S. Joseph was the group director of the fluid components group (FCG). However, Joseph had to leave ISRO and India in 2000 following his voluntary retirement taken on personal grounds. Subsequently, C. Amarasekaran and M. Yagnanarayana led the components design activities, and Baby T. Mathew managed the components production and assembly. In April 2000, Sridharan Dhas organized an internal seminar at Valiamala on 'INSAT-3B Propulsion Systems', where papers highlighting development issues and the flight performance of INSAT-3B propulsion system engines and fluid components were presented. In the same month, he retired, leaving behind a legacy of great achievements.

After this, the components activity came under M.K.G. Nair, who was also deputy director for Liquid Propulsion Stages (LPSs). In 2001, the components team faced a major material issue related to the stress corrosion cracking of PS2/GS2/L40 engine parts and stage fluid system components made from aluminium alloy 7020, even in unstressed storage conditions. A national committee discussed the problem and recommended the use of aluminium alloy 7075 forgings alone instead of 7020. This decision resulted in the large-scale rejection of the aluminium alloy 7020 and a subsequent delay in components production schedules. The components and modules performed well in GSLV flights in 2001 and 2002.

Meanwhile, the components for indigenous CUS engine systems developed by C. Amarasekaran and his team demonstrated satisfactory performance in the A1, A2 and A3 engine qualification tests from 2002 to 2004. The CUS system components developed by Yagnanarayana and his team also worked well in stage propellant expulsion tests. A. Manimaran, G. Sundaravadivelu, Poly P. John, S. Vinodkumar, R. Srinivasan, M. Radhakrishnan, B. Satishkumar and S. Murugan played commendable roles in the development of CUS engine and stage system components and modules. The ground qualification test of CUS for full flight duration was carried out successfully in 2007 with the satisfactory performance of components and modules.

After 2000, the demand for fluid components rose steeply, in view of the increase in frequency of PSLV/GSLV flights, as well as the new requirements for the development of L110/C25 stages for GSLV-Mk-III. Each GSLV-Mk-III vehicle required 181 fluid components for its liquid stages. The components production arrangement, introduced in 1993, could not catch up with this demand. While availability of fabricated parts was comfortable, the assembly of components reached a bottleneck. To overcome this problem, in addition to the efforts by CPP, 'end-to-end' production contracts were entered into with external industries, including MTAR Technologies and LTE, which were brought in to manufacture, integrate, test and supply 'ready-to-fly' modules and components. A. Alex and A. Manickavasagam worked as contract managers. In 2006, following the promotion of M.K.G. Nair as the associate director of LPSC, K. Ramajogi became the deputy director of Control Systems and Components (CSC). After his retirement in September 2006, C. Amarasekaran took over as the deputy director of CSC.

A. Manimaran and his team developed the components and modules for L110 stage and S200 FNC system of GSLV-Mk III. In 2009, the L110 stage hot test was successfully conducted with the satisfactory performance of components and modules. The same design engineers who worked for the CUS components were deployed for the C25 stage components development. In addition, D. Venkittaraman was also brought in from the SITVC team to support the development of crucial components for the C25 engine. Young and bright engineers, including S. Sunil, Benoy Joseph, V. Dilip, Raji George and Suryamani Tripathy, played a greater role in the development of C25 stage components. Meanwhile, additional manpower and infrastructure was provided from 2006 onwards, to support the new design and development efforts made towards the production of C25 stage and semi-cryogenic engine components and to strengthen the production wings.

The development and qualification of C25 stage components and modules were completed successfully with the involvement of the new team of engineers and technicians. Components function was normal in the tests of C25 engine subsystems and the main engine. From 2010 onwards, the development of fluid components for the 2000 kN semi-cryogenic engine was taken up in a big way, with D. Venkittaraman leading the activity. The semi-cryogenic engine system requires twenty-two components per engine. Today, the development of these components is in a very advanced stage. Recently, the components team successfully developed propellant flow control valves for the 800N engines with throttle, for the soft landing of Chandrayaan-2 Lander spacecraft on lunar surface.

In the last five decades, fluid components technology has evolved by leaps and bounds, starting from simple solenoid valves and moving on to high technology multi-functional modules and intricate spacecraft propulsion components. To date, thousands of these devices have been flown in ISRO's launch vehicles to help in the accomplishment of various missions. Orbit-manoeuvring thruster valves continue to function satisfactorily in about thirty spacecrafts orbiting the earth today. It is heartening to note that the present team of control systems and components led by M. Radhakrishnan is carrying on the good work done by the pioneers, besides making great strides in the development of semi-cryogenic engine components and mastering new technology areas such as propellant throttle valves for the Chandrayan-2 Lander engines.

12
French, German and Russian Collaborations

THE INDIAN SPACE PROGRAMME started with solid propulsion systems, and subsequently additions of liquid and cryogenic propulsion systems were made. My major responsibility was to develop and supply propulsion systems to meet the stipulated requirements of the launch vehicles and satellites of ISRO. In addition to developing the technology and human resources and establishing the infrastructure and facilities, I had another crucial duty – instituting test and evaluation facilities along with the support essential to develop and qualify different propulsion systems and subsystems.

Propulsion Systems: Test and Evaluation Facilities

To accelerate the task of creating these facilities, I thought it would be wise to seek guidance from agencies abroad that had demonstrated capability in launch vehicle programmes. I pitched this idea to Dr Sarabhai, and he agreed that ISRO should approach NASA. The response from NASA was initially encouraging and I could visit some of the propulsion test facilities in the US, hold discussions with the officials there and generate plans for collaboration. However, NASA refused such interactions and cancelled all activities immediately after India's first successful nuclear explosion test on 18 May 1974 at the Pokhran Test Range.

I recall that sometime in 1969 ISRO had entered into MOUs (Memoranda of Understanding) for collaboration in space activities at the international level with the Centre Nationale d'Etudes Spatiales (CNES) of France (the French Space Agency) and the Deutsches Zentrum für Luft-und Raumfahrt (DFVLR) of Germany (the German Aerospace Centre). Dr Sarabhai had appointed me as the ISRO coordinator for these two MOUs. I decided to approach SEP, France, through CNES and DFVLR for guidance and assistance to establish the facilities at STEX.

During 1970–71, Dr Satish Dhawan, the director of IISc and member of the Space Commission, was on sabbatical at the California Institute of Technology, Pasadena, USA. I met Dr Dhawan as per Dr Sarabhai's directions. We visited the Rocketdyne Engineering Corporation and the Marquardt Engineering Company, two aerospace firms in California. We saw some of the liquid propulsion test facilities and held discussions with their officials. In addition, I visited Aerojet General Corporation and Arnold Engineering Development Center (AEDC), Tullahoma, and saw some of their test facilities. Based on these visits, discussions with experts and Dr Dhawan's guidelines, I evolved the following strategy and implemented it to provide propulsion test facilities at STEX, initially for the SLV-3 Project.

(i) STEX would generate a document on propulsion systems test requirement, in consultation with the SLV-3 Project.

(ii) This document shall be discussed with the agency that is the consultant and technology provider. Based on the discussions, a document on test facilities requirement will be generated. The test facilities would carry out all the required tests on propulsion systems during the development and qualification phases.

(iii) Following discussions with the consultant, STEX was to generate a test facilities specification document, which would then be reviewed and approved by the consultant.

(iv) STEX would generate detailed design documents for all the test facilities with the layout, configuration, detailed designs of items to be manufactured, specifications of bought-out items and so on. The documents for assembly, testing, quality assurance and management system would also be included, to be reviewed and approved by the consultant.

(v) STEX would manufacture all the hardware with the assistance of external agencies, procure bought-out items, and establish the necessary assembly, integration and support facilities as per the approval of the consultant.

(vi) Consultancy and engineering services to establish and commission the facilities would be available from the consultant on mutually agreed terms.

French Connections: Sea-Level Tests for Solid Motors at STEX

In 1968, Dr Sarabhai founded STEX as a project at Sriharikota and nominated me as the project engineer. I was guided by a STEX project management board, which worked under my chairmanship and helped me monitor, review, guide and approve all activities during project execution. The mandate of STEX was to establish assembly and test facilities along with the necessary support amenities to test and qualify all four solid motors of SLV-3. At STEX, I created two project teams – the first one, under the leadership of M. Annamalai, to establish sea-level test and associated facilities, and the second, headed by Dr T.M.T. Nicolas, to institute altitude simulation tests and associated infrastructure.

SEP, France, had created and operated sea-level test facilities for the French solid motor programme at their plant at Bordeaux, but it did not have the facilities for High-Altitude Simulation Test (HAST). Under the guidance of T.N. Seshan, the joint secretary of DOS, and with the assistance of CNES, I negotiated with SEP, which was represented by R. Morin, the then commercial director. We concluded an agreement for technical collaboration between STEX and SEP in 1974 with a view to institute static test facilities for SLV-3. I formed a project team led by M. Annamalai at STEX. The project team of seven STEX personnel was deputed to SEP for a period of twenty-four weeks. The other members were K. Viswanathan, B. Raman, G.K. Mani, N. Satyanarayana, G.N. Pereira and Bosco. This team consisted of engineers, a draughtsman and a French translator.

SEP nominated their project team, with Tharrael as leader. They arranged group accommodation for the STEX team in a neighbouring housing complex with single rooms and all essential facilities. Since the STEX project team stayed together with their leader, they could work for extended hours beyond the SEP office time. The team implemented the foreseen strategy for consultancy, guidance and technology transfer and generated all the documents required for the following facilities:

(i) Sea-level static test facilities for performance evaluation and qualification of all solid propulsion systems of SLV-3. This process included: (a) Single-component measurement test stand for the measurement of the axial thrust of the solid motor; (b) Six-component measurement test stand for the measurement of side thrust and roll of the solid motor during SITVC system tests, in addition to axial thrust; and (c) Spin test stand to evaluate the performance of the spinning upper stage solid motor. It is noted that the fourth stage of SLV-3 is spin stabilized in flight.

(ii) Environmental test facilities for the qualification of the solid propulsion systems in simulated flight environment. This step included: (a) Vibration test facility for qualifying the upper stages and igniters of SLV-3 for simulated flight 'vibration' environment; (b) Shock test facility for qualifying the upper stages and igniters of SLV-3 for simulated flight shock environment; (c) Static acceleration test facility (centrifuge) for qualifying the upper stages and igniters of SLV-3 for simulated static acceleration experienced by the system in flight; and (d) Thermal/ humidity test facility for simulated thermal cycling and humidity cycling of the upper stages and igniters of SLV-3 to qualify them for storage conditions.

(iii) Data Acquisition and Recording Centre to acquire various parameters, such as thrust, pressure, temperature, strain and vibration during the test firing of these propulsion systems.

(iv) Electronics calibration and maintenance facilities for the calibration of various sensors and transducers; in addition, maintaining the various electronic equipment used in these facilities.

In addition to the joint designs and documents of these facilities, the STEX team generated the designs and documents for the following additional facilities on their own:

a) Solid stage assembly facility,
b) Hardware inspection facility,
c) Subsystem assembly facility,
d) Solid stage storage facility, and,
e) Material handling and technical services.

The STEX team returned to Sriharikota with all the designs and documents as intended and generated a detailed plan of action to establish these facilities in India. The development plan was reviewed and approved by the STEX project management board and implemented. The

STEX team established and commissioned these facilities in association with external industries to match the SLV-3 schedules. The first static test at STEX with minimal facilities was carried out in July 1974 on the ITS-1* proof motor. All systems performed well as per design, and it was a memorable occasion in the history of STEX.

Officials from SEP, Bordeaux, were invited to STEX and shown the facilities and operations. They were highly impressed by the manner in which the humble beginning made at SEP had grown into a world-class facility in 1979. They congratulated the STEX team for their excellent performance. Based on the knowledge, expertise and experience, STEX provided additional facilities and successfully implemented the test requirements of all solid motors for SLV-3, ASLV, PSLV and GSLV among others. Following the successful commissioning of STEX and the initial tests of solid motors for SLV-3, I handed over my responsibility to Retnaraj Jeyamani, the principal technical officer who had assisted me in all technical and managerial matters of STEX. I continued as an adviser of STEX for a few more years.

First Static Test of S125, First Stage Motor of PSLV

Though I have participated in a number of static tests of solid motors at STEX, I can never forget the thrill during the first test of the PSLV first stage motor with 125 tonnes of solid propellant. Rajaram Nagappa, head of the solid propulsion division of my group, led the development of the biggest solid motor at ISRO. A team of ISRO experts under my chairmanship reviewed and guided the process periodically for more than six months. The solid motor, which was

---

* The name given by STEX.

hitherto ISRO's largest, was assembled in the STEX test bed. At the time, I was away in Germany for a review meeting with Linde plc, the supplier of liquid hydrogen plant to ISRO, on the readiness and schedule for the supply.

The first test was scheduled for 21 October 1989. It happened to be the date of my return to India from Germany. Rajaram called me and expressed the desire of the PSLV project and the solid propulsion division for my presence at the test, along with the chairman of ISRO, the director of the VSSC, the project director of the PSLV and other senior officials. Accordingly, I changed my travel plans and flew from Frankfurt to New Delhi in the early hours on the day of the static test. I then took the first morning flight from Delhi to Chennai. Nagappa had arranged for a vehicle to pick me up at the Chennai airport and for an engineer from STEX to receive me. I drove directly to STEX and reached the security gate barely ten minutes before the scheduled test.

By special arrangement, the security gate was opened for me to proceed and I reached the block house of the test bed about five minutes prior to the test. The door of the block house was closed for security reasons, but I was let in with special permission. Getting inside, I stood close to the door at the back. In about ten seconds, the motor was ignited and I watched the successful burning of the largest solid motor till the burnout. Witnessing the first test of the first stage motor of the PSLV, along with the VIPs present there, was a thrilling experience. Immediately after the test, I walked down the central aisle and met the ISRO officials. It was a memorable milestone in the history of STEX and the PSLV project. I was extremely happy that the test was a grand success in its first attempt, despite the many complex issues we had faced during the development.

German Connection: High-Altitude Simulation Test (HAST) Facilities at STEX

The propulsion systems operating at high altitudes often experienced ignition failures. Such systems with large expansion ratio nozzles, which flow full at high altitudes, could not be evaluated with sea-level tests due to flow separation. Hence, I initiated studies on high-altitude ignition failures and the performance of propulsion systems at high altitudes during 1971–72. I recognized the merit in collaborating with overseas agencies that had successfully developed HAST facilities for testing and qualifying the upper stages of launch vehicles on the ground.

During the 1960s, the USA and the USSR were making rapid progress in space. Soon, Europe established the European Launcher Development Organization (ELDO) to develop its own launch vehicle 'Europa' with the active participation of the UK, France, Germany and Italy. The third stage of the three-stage Europa was a cryogenic stage, developed by MBB of Germany. All the facilities required for the cryogenic engine test at simulated high altitude and also for qualification testing of the integrated stage were established at the DFVLR test centre in Lampoldshausen. The Institute of Gas Dynamics under Stuttgart University, headed by Dr Dadieu, undertook the responsibility of establishing the High-Altitude Test (HAT) facilities needed to evaluate the performance of the cryogenic third stage.

This establishment was built in Lampoldshausen under the guidance of Dr Mueller, Dr Toofan and Dr Fabien. In addition, there were assembly, storage and test facilities for smaller-thrust liquid engines. Unfortunately, the ELDO programme was abandoned after eleven successive failures, and Ariane was established in 1980 as a multinational company by bringing together all the resources of Europe. I approached Dr Dadieu for guidance and assistance in establishing HAT facilities for ISRO programmes. His response was positive.

I also approached Dr Goetherdt, dean of the University of Tennessee's Space Institute in the US. He had been with Wernher von Braunin, establishing rocket test facilities in the 1940s, and later worked as chief scientist to build the aerospace facilities at the Arnold Engineering Development Center (AEDC), Tennessee. During one of his earlier visits to India, Dr Goetherdt agreed to provide training for ISRO scientists in his institute. Initially, he permitted Indian scientists to work at his institute to acquire the technology elements along with other United States Air Force officials who were part of the AEDC facilities. As part of the postgraduate programme in rocket propulsion, ISRO scientist Dr T.M.T. Nicolas could thus get very good exposure in the theoretical aspects of the HAT facilities. However, due to a ban imposed by the US government, no further help was possible from them in this regard.

In line with the programme with SEP for the sea-level testing of the SLV-3 propulsion systems, I concluded an agreement with DFVLR in 1973 to establish HAT facilities. Accordingly, a team of eleven STEX personnel were deputed to DFVLR, initially for a period of twenty-six weeks. The group consisted of Dr T.M.T. Nicolas who was the project leader, K.L. Valliappan, Mohammed Muslim, F.P. Joshua, Narendra Kumar, T.C. Sekar, Nandakumar, Devassy, Chandran, Ravindran and Venkadappa, and there were engineers, a draughtsman and a German translator. They were deputed to acquire technology to equip them to be able to establish High-Altitude Simulation Test (HAST) facilities for solid and liquid propulsion systems in general and, in particular, design a facility to test the solid propulsion fourth stage of SLV-3.

The STEX team was guided and assisted by a DFVLR team led by Dr Mueller in association with Dr Toofan, Dr Fabien, Dr Kenntner, Dr Shoen Harr, Dr Porsche and others. While designing the HAT facilities and availing of the training in DFVLR facilities, the STEX team was provided with food and accommodation in the DFVLR guest house on the campus. The guest house was well furnished and had a conference hall, which was effectively used by the STEX team as an office for discussions and review meetings. The configuration, layout and designs of the test facilities were finalized by the STEX and DFVLR teams.

The STEX team had an excellent working relationship with the DFVLR team, both professionally and personally. The DFVLR team was happy to have such an opportunity to guide and assist STEX to build high-altitude simulation test facilities and to be able to put their expertise and knowledge to use. Dr Fabien, the oldest member of the DFVLR team, used to say: 'Where else could DFVLR build such HAST facilities except in India!' The STEX team was exposed to technology elements, such as vacuum system, diffuser ejector system, barometric well concept, engine cooling system, steam generators using rocket fuels, and instrumentation and control system. The members designed various subsystems under the guidance of the German scientists and verified the design with the data available at DFVLR.

This procedure gave us the confidence to prepare specification documents and build high-altitude test facilities for the Indian Space Programme. During the design of the

HAST facilities, every member of the STEX team worked in different groups of DFVLR and established a good relationship with their German counterparts in the respective departments. Invariably, the STEX team had evening tea with the Germans and their family members, who also worked at DFVLR. On weekends, the Germans took STEX engineers to their farmhouses to experience life in the German countryside. During holidays, the Germans escorted them to the city halls, schools, parks, colleges and churches to help the visitors decipher the historical background of the nation. The shared social experiences and feelings added significantly to the technical knowledge acquired. This strong rapport continued even after our personnel returned to India.

After studying the ISRO rocket testing requirements, the following project reports were prepared by the HAST team in DFVLR, which then got reviewed by German scientists under my guidance:

1. HAT1 – for qualifying the third stage of SLV-3, using vacuum pumps, supersonic diffuser and steam ejectors;
2. HAT2 – for qualifying the fourth stage of SLV-3, which was used as the apogee motor of the satellite, using vacuum pumps, supersonic diffuser and steam ejectors;
3. HAT3 – for qualifying the control thrusters of SLV-3 and satellite RCS thrusters using vacuum chamber and vacuum pumps; and
4. HAT4 – for qualifying the liquid upper stage and engine with 3.0 tonne class engines, using vacuum chamber, vacuum pumps, supersonic diffusers and steam ejectors.

During high-altitude testing, low pressure was maintained around the rocket engine nozzle exit so that full nozzle flow was maintained without any separation, thereby ensuring that the engine developed full thrust as in operating altitudes. The supersonic diffuser, a simple gas dynamic device, enabled large volumes of hot exhaust gases from the rocket engine at 3,000K to be pumped across to create a thousand-fold increase in pressure. This process was new to STEX and a technology breakthrough was made by our team through this collaboration with DFVLR. The STEX team acquired the technologies to generate back pressure at the nozzle exit of the propulsion system by using supersonic diffusers, ejectors, vacuum pumping systems and so on. This process included the design concept of barometric seals and wells to discharge large quantities of liquids at low pressure into the atmosphere.

In addition to the supersonic diffuser with the test engine, multistage ejectors were required, depending upon the area ratio of the nozzle. These ejectors used steam as the driving fluid. The steam was produced in an unconventional method, using rocket chambers (0.5 m diameter x 1 m length) as steam generators. The steam was produced by injecting water into the combustion chamber of a small-thrust liquid engine with RFNA as the oxidizer and kerosene as fuel. The STEX team had generated designs for a 30 kg/sec flow of steam, which was operational at Lampoldshausen. It may be noted that this rate of steam flow was around that in a 100 MW thermal power station. When the STEX team built the HAT2 facility for the SLV-3 third stage at SHAR, the steam requirement was high. Hence, the 30 kg/see steam generator was upgraded to 37 kg/sec. The STEX team faced some combustion instability problems, which were solved subsequently by redesigning the combustion chamber.

The STEX team generated all the documents required for establishing the vacuum system, diffuser ejector system, large- flow stream generator system, instrumentation and control system and so on. The STEX team returned to Sriharikota in 1974 and started work on the HAST facilities one by one. Meanwhile, another sixteen weeks of operational training was arranged for a three-member STEX team that consisted of Dr T.M.T. Nicolas, Nandakumar and T.C. Sekar to participate in the Ariane cryogenic engine test programme in DFVLR.

The STEX team worked out a detailed plan of action, which was approved by the STEX project management board, and established the following facilities:

1. The HAT1 facility was established in Sriharikota during 1977. The SLV-3 third stage motor was successfully tested and qualified in that facility.
2. The HAT2 facility was also established along with HAT1 in Sriharikota in 1977. The SLV-3 fourth stage motor was successfully tested and qualified in this facility. The APPLE ABM was also tested and qualified in that facility in 1979.
3. Two HAT3 facilities were built, one in TERLS and the other in Bengaluru, in 1979. All qualification tests were conducted on the control thrusters used in launch vehicles and satellite programmes. These facilities enabled the testing of ignition of the control thrusters and spacecraft thrusters under deep vacuum conditions. The tests provided very valuable inputs

on finalizing the start sequence of these thrusters and the precise measurement of the thrust developed at vacuum.

4. The HAT4 facility was built in Mahendragiri in 1984. The fourth stage of the PSLV was tested and qualified in that facility. In addition, the LAM engine and control thrusters for the INSAT series of satellites were tested after necessary modifications. The fourth stage of PSLV used twin engines, each developing a thrust of approximately 750 kgf in vacuum. This engine had a regeneratively-cooled thrust chamber. The testing of this engine under high-attitude simulated condition validated the ignition in deep vacuum and the heat transfer characteristics. The diffuser–ejector system, which was basically a gas dynamic pump, was operated with air as the primary ejector fluid. The air stored at 300 bar in sixty 2.2m^3 high-pressure cylinders was regulated and used. This novel, simple and cost-effective system, used in the country for the first time, paved the way for the development of propulsion systems for launch vehicle upper stages and spacecrafts.

In the meantime, the Indian National Satellite (INSAT) was being developed. The propulsion systems to take the satellite from geo transfer orbit to geosynchronous orbit and for attitude and orbit control were developed by LPSC. The 440N and 22N bipropellant thrusters were designed and realized for this purpose. The HAST facility for testing the PS4 engine was augmented with a new diffuser–ejector system to test these thrusters, which resulted in a time-effective and cost-effective solution.

The HAT4 was a unique facility, because for the first time in the world, stored high-pressure air, instead of steam, was used as the driving fluid in such ejectors. I wish to express my appreciation and sincere thanks to Dr T.M.T. Nicolas and his team for successfully establishing the HAST facility and for testing the fourth stage of the PSLV at Mahendragiri in 1990. Commendable are the altitude simulation facilities and the test of the LAM engine used in the INSAT series of satellites at Mahendragiri in 1988. The successful testing of LAM for a duration of 4,000 seconds in a single-burn under-simulated high-altitude conditions is praiseworthy. This LAM was used successfully not only to raise the orbit of ISRO satellites but also in the first Chandrayaan mission.

The joint DFVLR–STEX project venture had an excellent understanding of the working system. During the deputation of the STEX team to DFVLR, two progress review meetings were organized at DFVLR. I participated in these meetings as the chairman of the STEX project management board, along with the director of DFVLR, the STEX team at DFVLR, and experts from DFVLR who were the technology providers. In addition, weekly review meetings conducted by Dr Nicholas, Dr Mueller, Dr Toofan and Dr Fabien, along with the joint project team, served to complete the design tasks on time.

Further, the STEX team had the opportunity to visit all the test facilities at DFVLR and to associate with their operations. Dr Nicholas and the STEX team maintained contact with their DFVLR counterparts even after their return to India. They also consulted DFVLR experts when the HAST facilities were instituted in India. I would like to place on record my gratitude to Professor Dadieu, Dr Mueller,

Dr Toofan and Dr Fabien for their professional guidance and unstinted support in establishing HAST facilities in India.

The Russian Connection: Cryogenic Test Facilities

In 1992, I visited the Research Institute of Mechanical Engineering (RIME), Salda, a Russian aerospace firm, and discussed a development plan to establish cryogenic systems assembly and test facilities at Mahendragiri. In 1993, I entered into an agreement for technical assistance. My visit to the ice-covered Salda facilities in the bitter cold of winter and the final discussions with Dr U.R. Rao and Dr Duneave while clinching the deal are still fresh in my mind.

The facilities to test the Cryogenic Upper Stage (CUS) propulsion system were established at Mahendragiri under the guidance of Russian experts. The CUS had a main engine that delivered 7,100 kgf thrust at vacuum and two steering engines each to deliver 200 kgf thrust. During the development of the Russian cryogenic engine (KVD-1), extensive tests were conducted at Russia at sea level as well as in high-altitude conditions. At Mahendragiri, the CUS engine with higher area ratio nozzle, designed for high-altitude operation, was tested at sea-level conditions in a novel way, using the Nozzle Protection System (NPS).

This system consists of a circular manifold mounted at the exit plane of the engine nozzle flush with the exit diameter, which feeds water and gaseous nitrogen at a specified pressure and flow rate, through slit-like openings circumferentially towards the axis. The mixture of water and nitrogen rises into the engine nozzle up to the plane of flow separation, thereby bringing down the thermal and vibration levels very low. In addition, the radial movement of the engine nozzle is restricted by shock absorbers all around. Thus, the high-altitude nozzle is tested at sea level. Of course, such a test proves the propulsion system performance at vacuum, except the thrust. However, the CUS engine was tested subsequently with nozzle at full flow in the HAST facility with a huge vacuum chamber and diffuser–ejector system to evaluate the vacuum ignition conditions.

Based on the training at facilities of Salda and the discussions with the Russian experts, the LPSC team gained adequate knowledge and came out with detailed procedures for the handling and servicing of liquid hydrogen at -253°C. The mechanical, instrumentation and safety systems for the cryogenic test facilities were very complex, and required meticulous planning, design, realization, installation and commissioning. It is heartening to note that the cryogenic test facilities for the CUS of the GSLV became operational with the first successful test of the CUS engine in early 2002. I express my sincere gratitude to D. Sarveson, S.M. Rajaratnam and their team for establishing and operating the complex cryogenic test facilities.

The infrastructure for the integration of cryogenic engine and stage were installed at Mahendragiri. The all-welded configuration of the cryogenic systems called for state-of-the-art Non-destructive Testing (NDT) techniques. Integration halls with clean room, insulation processing and surface treatment facilities were established. Integration procedures were jointly evolved by system designers and integrators. Cryogenic engines and stages assemblies are validated through extensive leak checks, by keeping the article inside the vacuum chamber.

The Russians wanted us to create this facility at the LPSC, and they proposed to supply the facility elements. Dr U.R.

Rao, the chairman of ISRO then, constituted an expert committee under my chairmanship to review the need for this facility. After assessing the merits and demerits, the committee came out with the recommendation to leak-check the engine assembly in the vacuum chamber and not the entire stage. This measure saved us around Rs 50 crore. I understand that our committee's recommendation is followed even today, thereby leading to the successful performance of cryogenic stages in flight. I appreciate the contributions of R. Krishnan Potti, Samuel Jayaprakash, J. Asir Packiaraj and K. Ramaseshan in this area.

Test facilities were also established at Mahendragiri for the cryogenic engine and stage subsystems qualification and acceptance. Turbine and pump tests, injector calibration tests and cryogenic temperature and fluids compatibility tests are worthy of mention. I appreciate the contributions of M. Annamalai, T.M.T. Nicolas, S. Murugesan, D. Sarveson, S.M. Rajaratnam, D.R. Selvanayagam, N.K. Gopalasamy, V. Udayanathan, J. Suresh Jeevan, K. Alaguvelu, L. Titus, B. Chellathurai and N. Selvamani in this connection. The programme to establish assembly and testing facilities along with support and service facilities was generated by the LPSC. It was reviewed in consultation with PSLV and GSLV and a detailed programme schedule was drafted to meet the project requirements.

For the success of the development of any propulsion system, establishing facilities for assembly, integration and testing in time to match the system development is critical. Dr Sarabhai had stressed the importance for sustainable and self-reliant technological developments. Adopting his guidelines regarding a complementary association of indigenous developments and simultaneous technology transfers from nations with the expertise required, I established the following:

1. Sea-level test facilities with technology transfer from SEP, Bordeaux, France, for solid propulsion systems;
2. Engine and stage test facilities with technology transfer from SEP, Vernon, France, for liquid propulsion systems; and
3. HAST facilities from DFVLR, Germany, for upper stage solid propulsion systems and upper stage liquid propulsion systems of launch vehicles, and control and orbit transfer liquid propulsion systems of satellites.

Over the years, our teams implemented the project with full confidence and also saved time and expenditure. I am happy to note that today ISRO has world-class propulsion test facilities for launch vehicles and satellites, comparable to any other spacefaring nation on earth. This achievement owes itself to a dedicated leadership and the tireless efforts of the entire workforce.

13

A World-Class Propulsion Test Centre: Mahendragiri

WHEN ISRO DECIDED ON the development of liquid propulsion stages, the need for a new site with adequate area and favourable terrain to establish large test stands and supporting facilities became evident. Dr Satish Dhawan, the chairman of ISRO at the time, assigned me the task of identifying a suitable site.

I approached the state governments of Kerala, Karnataka, Odisha, Maharashtra, Andhra Pradesh and Tamil Nadu, interacted with the officials concerned, explained our programme and indicated the requirements. The officials arranged field visits; some of my colleagues worked closely with me in visiting the sites, gathering maps, obtaining details and participating in the discussions with the authorities. We critically analysed every site to assess its suitability for our purposes.

In Kerala, I interacted with the district authorities in Thiruvananthapuram and visited the reserve forest areas in the Kottoor and Achankoil regions. The chief secretary of Karnataka, on the other hand, regretted that he could not allow the use of explosives, which was part of our requirements. In Odisha and Maharashtra, I spoke to the Hindustan Aeronautics Ltd (HAL) officials and examined the regions around the factories at Koraput in Odisha and Nasik in Maharashtra. In Andhra Pradesh, I communicated with officials of the Department of Industries and Commerce, and visited hill areas around Tirupati. Voluminous reports regarding the sites were generated under my guidance.

The chairman of ISRO stressed the need for further studies and asked us to conduct an Environment Impact Assessment (EIA), to address the repercussions of the activities on land, water and air during normal test operations, and in the event of unforeseen failure. Dr Dhawan also arranged for me to meet Dr Madhav D. Gadgil, then professor at IISc, Bengaluru. My colleagues consulted experts independently.

Based on our studies and discussions with specialists in various disciplines over three years, we came up with an exhaustive report. It addressed several parameters, including environmental impact, proximity to the ISRO establishment, and infrastructure and logistic facilities nearby and made specific recommendations. I remember the sincere and valuable inputs of S. Nambinarayanan, T.P. Rangamani, A. Chandran, K. Manikantan, D. Subramanian, K.S. Pillai, M.K. Narendranath, N.K. Gopalasamy, Dr R. Natarajan and their associates in the successful completion of the task. I also thank R.D. John, chief engineer in DOS, and some of his colleagues who joined us in the endeavour.

The chairman asked me to prepare a comparative analysis of all the sites suggested by the state governments. Our team identified a number of parameters, compared their relevance and importance and assigned weightage to each aspect. Using the data and information available, marks were assigned based on every parameter to the sites. The exercise gave us a fair assessment of the various options. Our report was reviewed critically by a team of experts, including R.D. John and P. Sudarson from the ISRO headquarters in Bengaluru.

Finally, ISRO zeroed in on Mahendragiri (Photo No. 51), which was located along the border shared by the Kanyakumari and Tirunelveli districts of Tamil Nadu. I held meetings with the district officials and T. Thiravium, an Indian Administrative Service officer serving as the chief secretary of the Government of Tamil Nadu then, who conveyed the governmental approval of the assignment of the 7,000 acre land in Mahendragiri Hills on a very long lease. The district officials were quite helpful in the acquisition and transfer of private lands that were within our safety zone too. This step was the starting point for the major liquid propulsion test facilities at Mahendragiri.

Liquid Propulsion Test Facilities – A Unit of LPP

In 1982, the Liquid Propulsion Test Facilities (LPTF) were established as a unit of the LPP at Mahendragiri, with R. Karunanidhi as the manager and myself as the director of LPP. The development of the site and the construction of buildings and other facilities were initiated in a phased manner.

The primary objective of LPTF was to establish assembly and test facilities for the Vikas engine, its subsystems and the support services required. Fortunately, the ISRO engineers who were deputed to SEP had sufficient exposure to the facilities in Vernon, France, which were used for the Viking engine development. The experience gained in the process helped us plan the following facilities initially at LPTF:

1. Test stands to test engine, stage, gas generator, turbopump and components. To provide the necessary support for test operations, the purchase of handling equipment, maintenance shop, safety devices, nitrogen plant, neutralizer plant, effluent treatment plant, cooling system for deflector, photography, common control and instrumentation room, and so on were planned. The infrastructure was upgraded in stages to cater to future needs as well.
2. Propellant storage and servicing facilities. Towards this purpose, separate bulk storage of fuel and oxidizer was envisaged with storage tanks of 35 cubic metre (cu.m) capacity each in open spaces with concrete floors. Propellant transport was provided with separate transfer tanks of 25 cu.m capacity. In addition, neutralization and disposal systems, safety monitoring and control systems, quality assurance laboratory, firefighting systems and so on were proposed.
3. Instrumentation and control room. This facility was to house an analog measurement system with transducers, signal conditioner, UV recorders, magnetic tape recorders and so on, for 175 measurement channels, twenty-six event monitoring and digital measurement systems with digital computer and peripherals with online data acquisition and offline data processing.

Moreover, a command system with automatic sequencer, threshold detectors, safety equipment and auxiliary equipment, such as CCTV and intercom, were to be kept inside.

4. Hydraulic test facilities for high flow calibration of injectors, flow metres, engine valves, components, etc., and for POGO and sloshing studies. Towards this purpose, water tanks with 30 cu.m capacity to supply water with a flow rate of 150 litres per second (lps) at around 1.5 bar and an air reservoir with a capacity of 1.23 cu.m at 250 bar to supply compressed air were suggested.

While the development of these facilities were in progress, the change in the configuration of the fourth stage of the PSLV in 1985 necessitated additional test facilities at Mahendragiri, with the capability to test the upper stage engine and stage under simulated high-altitude conditions on the ground. The following facilities were established at Mahendragiri to meet the PSLV assembly and testing requirements:

1. Vikas engine assembly facility
2. Pre-assembly workshop
3. Disassembly facility for tested Vikas engines
4. Sand blasting and coating facility for nozzle
5. PS2 stage integration facility
6. PS4 stage integration facility
7. Proof pressure test facility
8. Principal test stand for liquid engines and stages

9. Gas generator test stand
10. Auxiliary test stand
11. Instrumentation and control centre
12. PS4 engine and stage test stand
13. Thermal painting facility
14. High-pressure gas cylinders test facility
15. Fabrication facility
16. NDT facility (X-ray)
17. Metrology laboratory
18. Material handling and transportation division
19. Surface treatment facility
20. Surface cleaning facility
21. Material testing laboratory
22. High-pressure installation
23. Storage facilities for propellants UDMH, N_2O_4 and MMH
24. Storage command
25. Safety command
26. MON-3 processing plant
27. Effluent treatment plant
28. Meteorological facilities
29. Electrical sub-stations
30. Diesel generator power plants
31. General stores and raw material stores
32. Bonded stores
33. Engineering maintenance facility
34. Transducers calibration facility

35. Instrumentation and electronics laboratory
36. Technical photography division
37. Library and documentation
38. Ground-level reservoirs for process water
39. Administration, accounts and purchase
40. Security posts
41. Fire station
42. First-aid centre
43. Canteen
44. Guest house
45. Emergency staff housing colony and
46. CISF barracks and housing colony

In view of the hazardous nature of the activities, in addition to the introduction of appropriate safety systems, a full-fledged fire station was sanctioned at Mahendragiri under the safety command. It was supported by a first-aid centre equipped with the basic amenities manned by medical officers and a team. The material handling and transportation of over-dimensional rocket stages under vibration-free and shock-free conditions is always a big challenge. We successfully realized these systems, thanks to the diligence of engineers S.S. Frederick and Solomon Chandra.

ISRO then felt the need for a GSLV with CUS to boost the payload capability. In a GSLV, except for the core solid booster, the remaining six stages are liquid propulsion stages. The development of the liquid strap-on stages (L40 – four in number), a liquid second stage (GS2) and CUS were taken up by the LPSC. Vikas engines and their variants were used in all the five liquid stages. Stage engineering of the L40 and GS2 were almost similar to the second stage of PSLV (PS2).

I can recall quite clearly the day when we, along with R.D. John and his team, performed the Bhumi pooja (a ritual blessing the land) for LPTF at Mahendragiri in 1982. A huge scorpion and a snake showed up in their attempts to escape at two spots at the recently cleared site. Today, the jungle has been transformed into a world-class facility for the Indian aerospace programme. The magnificent change is the result of the untiring work of my colleagues. I would like to express my appreciation for R. Karunanidhi and his team. V. Shanmugavelu, the deputy manager, became the manager when Karunanidhi left ISRO. S. Murugesan was responsible for the control and instrumentation activities. D. Sarveson managed liquid propellants and related activities. S. Lakshiminarayanan looked after test stands and assembly activities in the initial stages. S. M. Rajaratnam, R. Krishnan Potti and M. Parasupillai took over when Lakshiminarayanan quit ISRO. I also acknowledge the immense support given by R.D. John and his team in the construction activities and in establishing the infrastructure.

As indicated earlier, the vast expanse of Mahendragiri was a wilderness when we occupied it. During the months of June and July, heavy winds caused soil erosion. To control this loss, we planted thousands of saplings, which grew well given the fertile soil and formed a well-balanced landscape of flora and fauna. I specially thank S. Thankaraja, our engineer, and T. Thiraviam, our assistant, for their very sincere efforts in nurturing the land at Mahendragiri. Their enthusiastic attempts to raise a colony of peacocks in the Mahendragiri forests, by bringing in nearly 150 birds from Sathur region with due permission from the

Ministry of Environment and Forests (MoEF), Government of India, deserve special mention. Initially these birds were kept inside a properly fenced area near the canteen, fed with millets, and reared; after about a month, they were released into the open forest. The peacocks not only added to the beauty of the neighbourhood but also prevented the spread of poisonous reptiles in the area, thereby facilitating undisturbed test operations.

The officials at Mahendragiri had to travel to the airport and the ISRO establishments at Thiruvananthapuram frequently. Nagercoil town was suitable for their residential purposes. As the town had public amenities such as schools, colleges, hospitals, shopping facilities, rail and road connectivity, and entertainment centres, we preferred to establish the staff housing colony there. The Roman Catholic Diocese of Kottar was kind enough to give us adequate land near Simon Nagar, in the middle of the town. Today, I am happy to note that about 300 ISRO families are comfortably settled there. Later, another housing colony was established in the neighbouring village of Valliyoor.

Over the years, the LPSC at Mahendragiri developed considerably under the able leadership of M. Annamalai, L. Muthu, Mohammed Muslim, V. Ganesan, D.R. Selvanayagam, D. Karthikesan, S. Rakesh, Dr P.V. Venkatakrishnan, S. Pandian, T. Mookiah and K. Alaguvelu among others. I congratulate all of them for their vital work in the liquid propulsion programmes of ISRO. Today, Mahendragiri boasts of efficient capabilities and facilities to assemble, test and qualify earth-storable and cryogenic liquid propulsion systems and their subsystems, with a wide range of thrust levels both at sea level and at simulated altitude conditions. The place has state-of-the-art instrumentation and control facilities along with all the support infrastructure necessary (Photo Nos 52, 53 and 54).

Today, Mahendragiri stands tall, balancing perennial technical activities and security with ecological conservation. Whenever I travel along the national highway between Kavalkinaru and Panagudi, I am pleased to see the world-class propulsion test centre in the ISRO Propulsion Complex and feel proud that I laid its foundation and contributed to its development from the early 1980s to 1995.

51. Mahendragiri site

52. Principal test stand to test high thrust liquid engine, stage and turbo pumps.PSLV second stage is under test

53. Twin Test Bay for sea-level test and high altitude simulation test with air jets and supersonic diffuser to test and qualify PSLV fourth stage, INSAT Liquid Apogee Motor, INSAT AOCS thrusters

54. Static test of twin engines of PSLV fourth stage

14

Launch Vehicles of ISRO: SLV-3, ASLV, PSLV

THE SPACE COMMISSION OF India guides and approves the space programme of the nation. It is executed by the Department of Space (DOS) through ISRO under the overall responsibility of the secretary who is also the chairman of the Space Commission and ISRO. The mandate of the R&D centres of ISRO is to develop technologies, establish infrastructure, recruit manpower and contribute to the launch vehicle and satellite programmes. In this chapter, I would like to present the propulsion systems developed and supplied to SLV-3, ASLV and PSLV, the first three launch vehicle projects of ISRO.

SLV-3: The First Launch Vehicle of ISRO

SLV-3 was the first launch vehicle of ISRO (Photo No. 55), developed by the VSSC, with A.P.J. Abdul Kalam as the project director. Around 1970, Dr Sarabhai wanted to initiate the first launch vehicle project of the nation on a fast-track basis, to put a modest satellite in a low earth orbit (LEO) in five years. He was aware that ISRO or any other agency in India did not possess the experience, expertise or infrastructure required to take up such an arduous task then. ISRO had just started work on solid propulsion. Hence, a solid propulsion system was the only option for the quick development of a launch vehicle.

The Evolution of SLV-3

Having established the technology, infrastructure and human resources to develop and supply sounding rockets, Dr Sarabhai initiated the next phase of the Indian Space Programme – ISRO Launch Vehicles. He reviewed the history of the development of launch vehicles around the globe and, based on it, selected the baseline specification of the first launch vehicle to be developed in order to launch a 35 kg satellite in a nominal 400 km LEO with a100-day lifespan. He wanted the first launch vehicle to be simple and operational, with minimum augmentation of infrastructure and expenditure.

Dr Sarabhai focused on the configuration of the launch vehicle and the management system. He wanted the configuration to be similar to that of any other operational launch vehicle elsewhere in the world in order to ensure its feasibility. The vehicle was to include different types of control systems, for use in any launch vehicle. Developing it would ensure that ISRO established the technologies with infrastructure to meet future launch vehicle programme needs. On 23 April 1968, Dr Sarabhai constituted a satellite feasibility study project, with Dr Y. Janardana Rao and P.P. Kale as conveners. The committee submitted its report in November 1968 with four different launch vehicle capabilities to put a 20–50 kg satellite into a nominal 400 km circular orbit. Dr Sarabhai chose the third configuration, SLV-3, which was similar to NASA's operational SCOUT, a four-stage vehicle with all the four stages powered by solid motors.

Regarding the management system for the Indian launch vehicle, Dr Sarabhai explored viable options, taking into consideration the status of technology and human resources available. It was around this time that the European nations were in the process of developing a major launch vehicle named Europa under the European Launcher Development Organization (ELDO). Europa was a three-stage vehicle – with the UK developing the semi-cryogenic first stage with the LOX-RP propellant made by Rolls Royce, France the earth-storable second stage with N_2O_4-UDMH (Nitrogen tetroxide-Unsymmetrical dimethylhydrazine) propellant made by SEP, and Germany the cryogenic stage with LOX-LH_2 propellant made by MBB and many others for the remaining subsystems. This European venture pooled the resources of many countries and simultaneously provided autonomy in the developments.

The coordination and management of the three major nations was very complex. Though ELDO was ultimately unsuccessful, the principles behind it gave a new dimension to the management system for launch vehicle projects, which required multidisciplinary technology inputs and optimum utilization of technology, infrastructure and human resources from diverse contributors. Subsequent to the closure of ELDO, after eleven unsuccessful launches, the organization and management system were modified and a company called Arianespace was established in 1980. Arianespace, like its predecessor, successfully pilots the European launch vehicle developments with the participation of leading aerospace agencies on the continent. ISRO had established a laudable association with aerospace agencies in France regarding the sounding rocket programme more than thirty-five years ago.

In April 1969, Dr Sarabhai introduced an innovative concept to carry out the detailed design and optimization studies of the launch vehicle. He established seven design projects with project teams and leaders, which were interdependent as well as independent. They were the following:

(i) DPS-1 (Design Project Stage-1) for the first stage, with Dr V.R. Gowariker as the project leader;
(ii) DPS-2 for the second stage, with M.R. Kurup as the project leader;
(iii) DPS-3 for the third stage, with myself as the project leader;
(iv) DPS-4 for the fourth stage, with A.P.J. Abdul Kalam as the project leader;
(v) DPGP (Design Project Guidance Package) for control and guidance, with Dr S.C. Gupta as the project leader;
(vi) DPE (Design Project for Electronics), with P.P. Kale and B. Ramakrishna Rao as the project leaders; and
(vii) DPLV (Design Project for Launch Vehicles) for the overall vehicle, with Dr Y.J. Rao as the project leader.

Each project team was to carry out the detailed design and analysis, generate the development plan, define the role and responsibilities of the participating agencies, and decide the milestones, schedule, budget, human resource requirement, quality assurance plan and management system among other things. The project developments were to be undertaken after the completion of the review and approval. It occurred to me that Dr Sarabhai was planning a modified ELDO management concept to provide opportunities for a few personnel and bring out their leadership skills and demonstrate innovative concepts.

It must be noted that ISRO did not have personnel with prior expertise and background in launch vehicle development. Dr Sarabhai's approach triggered the imaginations of a number of potential leaders and specialists who came out with original ideas to resolve the complexities in technology and management. These people were wedded to the vision and mission of Dr Sarabhai and united solely by his vision. I was selected to be one among the seven project leaders for the third stage (DPS-3) of SLV-3. I formed the project team by redeploying some of my colleagues and recruiting new ones. With earnest effort, we initiated the design activities and interfaced with all the project teams and external agencies that could guide us towards success and support us.

As the director of the propulsion group of the VSSC, my responsibility was to develop and supply all four solid propulsion motors, and also the SITVC and RCS, for SLV-3. In addition to the development of these propulsion systems, establishing the requisite ground test facilities was a major task. Our experience was limited to the much smaller sounding rockets. None of us in the propulsion group had any experience in simulating flight conditions for ground tests. During this period, ISRO had planned to institute large propellant manufacturing plants and a launch complex with facilities for assembly, integration, checkout and so on, at the Sriharikota centre (SHAR), where vast areas of land were available.

Considering these factors and the safety requirements, ISRO decided to establish the facilities for testing the solid propulsion systems at Sriharikota. As indicated earlier, the Static Test and Evaluation Complex (STEX) was created by Dr Sarabhai at Sriharikota with me responsible as the project engineer in 1968. It provided assembly and test facilities and carried out the development and qualification programmes of all solid propulsion systems of SLV-3.

Regrettably, the planned design projects could not be completed due to the sudden demise of Dr Sarabhai on 30 December 1971. Dr M.G.K. Menon assumed the interim charge of chairman in January 1972 and held it till September, when Dr Satish Dhawan, the director of IISc until then, took charge of the newly constituted Department of Space (DOS) and the Space Commission. He was the secretary of the DOS, the chairman of the Space Commission and the chairman of ISRO. He reorganized the SLV-3 development project, with A.P.J. Abdul Kalam as the project director in 1972.

The Configuration and Development of SLV-3

Though a lot of information has been published on the SLV-3, I would like to briefly present some details for the readers to appreciate the quantum of efforts in multidisciplinary areas. The SLV-3 was a four-stage vehicle, powered by four solid rocket motors, with a total length of 17 m and lift-off weight of 17 tonnes. It could impart a velocity of 28,000 km/hr to a 35 kg satellite and place it into an elliptical orbit. It had forty-four major functional subsystems with four main constituents: (i) vehicle propulsion systems, (ii) rocket systems, (iii) guidance and control systems, and (iv) vehicle electronics systems. I will discuss them briefly below.

(i) The vehicle propulsion system of the SLV-3 had four solid propulsion system stages. The first stage was with a 1,000 mm diameter metallic motor case, fabricated from 15CDV6 steel sheets and forgings in three segments and with 8.6 tonnes, indigenously developed PBAN* solid propellant with an Isp of 254 seconds and burn of 50 seconds. The second stage was with 800 mm diameter metallic case with 15CDV6 and with 3 tonnes of monolithic PBAN propellant grain – with an Isp of 268 seconds and burn of 44 seconds. The higher specific impulse was due to the higher expansion ratio of the nozzle. The third and fourth stages were of 800 mm and 650 mm diameter, respectively, with fibre-reinforced plastic motor cases developed in-house, a 1 tonne high-energy propellant (HEF 20) in the third stage and 262 kg HEF 20 in the fourth stage. The Isp of the fourth stage was 284 seconds (vacuum). I had the overall responsibility of designing, developing and supplying all four solid propulsion stages in the propulsion group of the VSSC, in association with Project SLV-3 and other groups of the VSSC.

(ii) The rocket systems of the SLV-3 comprised (a) a stage separation system, (b) a destruction system, and (c) a heat shield. For the first two stages, the Flexible

---

* Polybutadiene acrylonitrile

Linear-Shaped Charge (FLSC) was located between the stages and used in the stage separation system. For the third and fourth stages, ball-type separation system was employed. FLSC-type destruction systems were housed in the first three stages of the vehicle, which would destruct the vehicle, if necessary, based on the ground command issued after factoring in safety considerations. A fibreglass honeycomb heat shield, provided around the satellite and the fourth stage, protected them from aerodynamic heating during atmospheric flight; it was separated from the vehicle when the vehicle crossed the dense atmosphere at about 85 km.

(iii) The guidance and control systems of the SLV-3 were responsible for three-axis stabilization and for steering the vehicle along a pre-set trajectory. The guidance system used a four-gimballed stabilized inertial platform to sense the vehicle attitude. The autopilot compared attitudes (the position x, y, z in space) with command angles as given by the pitch programme stored in the vehicle attitude programmer with launch references for yaw and roll axes. The error signals and the vehicle body rates by rate gyro package generated the command signals to the control systems. The SLV-3 had four different types of control systems:

(a) Secondary injection thrust vector control (SITVC) system with proportional injection valve and strontium perchlorate as injectant was used in the first stage during the first 17 seconds of the powered flight for pitch and yaw control.

(b) Fin tip control (electro-hydraulically operated aerodynamic control surfaces) was used for role control throughout the flight and for pitch and yaw control beyond 17 seconds.

(c) In the second stage, bipropellant (red-fuming nitric acid and hydrazine) on-off RCS was used for pitch, yaw and roll control during both powered phase and coast phase flight regimes. For pitch and yaw control during the powered phase, 250 kgf bipropellant thrusters were used, and 50 kgf bipropellant thrusters were used for pitch and yaw control during the coast phase. For the roll control in both powered phase and coast phase, 10 kgf bipropellant thrusters were employed.

(d) For the third stage, monopropellant (hydrazine with indigenously developed catalyst) on-off control (RCS) was used for pitch, yaw and roll control during both powered phase and coast phase flight regimes. During the powered phase, 25 kgf monopropellant thrusters were used for pitch and yaw control and 5 kgf monopropellant thrusters for pitch and yaw control. For the roll control during both powered phase and coast phase, 1 kgf monopropellant thrusters were used.

The fourth stage was spin-stabilized. The different control systems were evaluated for performance by ground tests. In addition to computer simulation, the total guidance and control system chain was tested in Hardware-in-Loop Simulation (HILS) tests. I had the overall responsibility of designing, developing and supplying the SITVC control system for the first stage, the bipropellant RCS for the second stage and the monopropellant RCS for the third stage in my propulsion group, in association with Project SLV-3 and other groups at the VSSC.

(iv) The vehicle electronics systems of SLV-3 consisted of telemetry, tele-command, tracking and sequencing systems. The telemetry system employed both FM/FM and PCM/FM schemes to monitor about 400 vehicle parameters in flight, including motor pressure, temperature, guidance commands and attitude errors among others. On-board tracking included a C-band transponder and tone range receiver. The vehicle sequencer generated the actuation commands for stage ignition, separation and so on. The equipment bay housed most of the guidance and electronic subsystems.

In 1972, Dr Satish Dhawan appointed the members of the SLV-3 project team, with A.P.J. Abdul Kalam as the project director. The team was responsible for the design, development and launch of the first Indian launch vehicle to put the Rohini satellite in LEO. The project director was fully supported by the chairman of ISRO, the director of the VSSC, officials at ISRO headquarters, and all the personnel in the various groups and divisions. The task team prepared a comprehensive management plan, which was drawn up on the basis of brainstorming sessions conducted during fifteen meetings of the Space Scientific Committee (SSC) of the VSSC under the chairmanship of its director. The plan was finally approved, and Dr Brahm Prakash, the director of the VSSC, issued a comprehensive office order, which was successfully implemented.

Project SLV-3 had a core team, which reported to A.P.J. Abdul Kalam. The development tasks were subcontracted to different groups and divisions of the VSSC and other ISRO centres, coordinated by the SLV core team and managed by the project director. National laboratories and industries in both public and private sectors were associated in the development in a major way. My contribution to the SLV-3 launch vehicle is summarized below:

(i) As a member of the project management board of SLV-3, I actively participated in planning, monitoring, reviewing, guiding and approving the overall development of SLV-3, with special contributions in the areas of propulsion and mechanical engineering.

(ii) As the director of the propulsion group in the VSSC, I contributed to the development of all the propulsion systems in the 1970s, leading to the successful launch of SLV-3. Though I had the overall responsibility of overseeing the development of solid propulsion systems, I made significant contributions in the areas of grain design and interior ballistic analysis, and the design and development of the nozzle and the ignition system. I also developed and supplied RCS and SITVC control systems for the first three stages of SLV-3.

(iii) As the chairman of the standing review committee for the SLV-3 solid rocket motor, I carried out the review of all the four stages of the SLV-3 – from design to the flight clearance of all the four launches of the vehicle.

(iv) As project engineer and chairman of the STEX project management board, I planned and established, in the early 1970s, the facilities to assemble and test SLV-3 motors. This would help evaluate the performance of the propulsion systems and conduct their environmental testing, which included vibration, thermal cycling, constant acceleration, spin and so on.

(v) I successfully organized and executed a collaborative programme with SEP to establish sea-level test facilities. Another collaborative programme with DFVLR, under the ISRO–DFVLR programme, provided HAST facilities for the fourth stage solid motor of SLV-3, the liquid propulsion control thrusters and the upper stage liquid engines.

(vi) After the commissioning of the sea-level test bay at STEX, I handed over the charge of the facility to Retnaraj Jeyamani, the principal technical officer of STEX at the time, and continued as a consultant. I participated in reviews related to the establishment, operation and maintenance of new facilities. I also actively participated in the establishment of the HAST facility for testing the fourth stage of SLV-3 at STEX.

The Maiden Flight of SLV-3

ISRO and millions of Indians had pinned their hopes on the maiden flight of SLV-3 on 10 August 1979. Hundreds of engineers and scientists had worked with devotion, and multifarious industries had contributed their might, towards the realization of this event. Sadly enough, it was only a partial success. Of the forty-six major subsystems, thirty-six performed well during flight and all launch complex facilities and ground stations were fully operational. The performance of the flight was normal during the first stage operation. When the second stage was ignited and it started operation, the vehicle went out of control and splashed into the Bay of Bengal, 317 seconds after take-off.

The bipropellant RCS, which provides control forces and moments in flight, could not deliver them to control the flight. It appears that all the RFNA in RCS had been depleted on the ground before take-off after the warm-up pulse, executed at T-8 minutes as per schedule. Hence, RCS did not have RFNA in flight and could not provide the control force. There were two possibilities for this oxidizer depletion. The first one was due to the solenoid valve in the RFNA circuit not closing fully because of some contamination; the second was on account of the fusing of the relay in the valve's electrical circuit during the warm-up pulse operation, which resulted in the valve being left open whenever the launch vehicle system went into internal power supply mode. As per the countdown sequence, the launch system goes into internal power supply mode at T-8 minutes, T-4 minutes and T-3.5 minutes to carry out the warm-up pulsing of the second stage RCS system (at T-8 minutes) and the third stage RCS system (at T-4 and T-3.5 minutes). Subsequently, the vehicle goes into internal power mode at T-3 minutes once and for all.

Project SLV-3 examined the situation through its failure analysis committees at different levels and concluded that the most likely cause of the malfunctioning of the solenoid valve was 'contamination'. The process of contamination and the basis of the conclusion were not explained. There was no reference to any alternative mode of failure. This is the official version that remains in the SLV-3 records. Along with K. Sivaramakrishnan, who was responsible for this RCS development, I analysed the data of the history of tank pressure on the oxidizer circuit from the time of warm-up pulse till launch. We were convinced that the most probable cause of the solenoid valve failure was the malfunction/fusing of the switching relay in the vehicle's electrical circuit supplied by the SLV project during the first warm-up pulse operation, which resulted in the valve being left open whenever the system went into internal power mode. Since the ISRO management had accepted the explanation that failure was due to contamination, our view was rejected even at the highest levels and remained unreported in the SLV-3 documents.

Sivaramakrishnan, some of my other colleagues and I felt the need to set right the failure analysis record of the first SLV-3 flight – for historians and the future generations. Accordingly, Sivaramakrishnan met Dr Dhawan on 5 December 1979 and explained our analysis of the RCS failure in the SLV-3 second stage. He presented the simulation studies in detail – first, with the postulation of fluid contamination causing the valve to remain partially open and second, with the postulation that the relay in the drive circuit of the pitch-up oxidizer valve had fused. After some discussion and clarifications, Dr Dhawan accepted our view. I shall reproduce the communication from Sivaramakrishnan for record:

> I had carried out simulation studies, one with the postulation of [the] SLV3 failure being caused by contamination resulting in the valve being kept in the partially open condition and the second, with the postulation that fusing the electrical relay of pith-up

oxidizer valve during the warm-up pulsing causing the failure of the mission.

Towards proving the first postulation, I introduced various levels of contamination up to 250 micron particle sizes and subjected the oxidizer valve for about 25,000 pulses but could not simulate the failure of valve in the partially open condition. For simulating the partially open flow condition of the oxidizer valve I introduced a 1 mm port diameter valve in the circuit and kept it on for eight minutes in the integrated Proto RCS package, but this test could not simulate the pressure history observed in the gas tank of [the] SLV-3 second stage RCS from T-8 minutes till launch.

For simulating the second postulation, [the] SLV3 second stage Proto RCS was integrated and [the] gas tank was pressurized to 220 bar. The valve was kept on as per the pre-launch sequence of operation corresponding to the duration whenever the system goes into on-board power. This simulation study could identically reproduce the pressure history in the gas tank of [the] second stage from T-8 minutes onwards. Both the above simulation studies were carried out in presence of SLV-3 representatives. Hence, my conclusion was that the failure of [the] SLV-3 maiden flight was caused by relay failure and not by contamination. I presented this data to the Dr Gupta committee but they did not concur with my conclusion.

I presented the above failure analysis results to Prof. Dhawan on 5 December 1979. After some discussions and clarifications, he accepted my findings. He asked me whether he should make any new press statements based on this new finding. I requested him not to do so, since the Project SLV-3 has already published the failure mode based on project assessment.

After the partially successful maiden flight of SLV-3, the team demonstrated its resilience and commitment by starting activities immediately, ending only with the successful SLV-3-E02 flight on 18 July 1980. The flight placed a 35 kg Rohini satellite into an LEO. It was a big step for India, which made us one of the spacefaring nations on the planet with satellite launch vehicle capability. Project SLV-3 was successfully completed with the launch of two more satellites. My team supplied all the components and subsystems for the four SLV-3 vehicles and participated in the assembly, testing and checkout operations and launch operations.

ASLV: The Second Launch Vehicle of ISRO

The Augmented Satellite Launch Vehicle (ASLV) (Photo No. 56) was the successor of the SLV-3. It had the ability to launch a nearly 150 kg satellite. It is a five-stage, all-solid launch vehicle, configured by adding two of the first stages of SLV-3 as strap-on 'zero-eth' stage on both yaw directions of the proven SLV-3. The configuration was simple and the subsystems of SLV-3 were used to minimize new subsystem developments. Consequently, the development period and expenditure were minimized. Thus, the ASLV configuration had two strap-on stages, ASO-1 and ASO-2, and the four stages of SLV-3, namely, AS-1, AS-2, AS-3 and AS-4, with a lift-off mass of 40 tonnes, total height of 24 m and payload capability of 150 kg in LEO. Project ASLV was the second launch vehicle development project of the VSSC, which proposed to demonstrate new additional launch vehicle technologies.

It was not planned as an operational launch vehicle for the ISRO Space Programme.

The ASLV was approved on 1 July 1982 with M.S.R. Dev, an aeronautical engineer who had worked closely on the SLV-3 as the ASLV project director. The new technologies proposed to be proved by the ASLV team for future applications were (i) strap-ons to increase the launch vehicle capability, (ii) closed loop guidance for precise injection of satellites in specified orbit, (iii) yaw manoeuvring required for many polar launch missions from Sriharikota, (iv) vertical launch vehicle integration, needed to assemble big launch vehicles, and (v) bulbous heat shield, needed to accommodate larger payloads.

The ASLV project team carried out four launches: the first on 24 March 1987, the second on 13 July 1988, the third on 10 May 1992 and the last one on 4 May 1994. The first two launches failed; the first was due to the non-ignition of AS-1. Following detailed studies, analyses and deliberations, the FAC concluded that the trouble had occurred consequent upon the random failure of a component associated with the ignition chain. The second setback was on account of the non-availability of control forces in flight. The FAC held that it was because the sequencing of flight events did not take care of the possible variations in the burning durations of strap-ons and the inadequacy of the autopilot, which was sluggish to respond. Consequently, the vehicle deviated from the planned trajectory. The aerodynamic forces on the vehicle became formidable and the vehicle broke up.

The FAC and the Expert Review Panel (ERP) analysed the failures thoroughly and made specific recommendations, which were subsequently implemented. The major remedial actions were to redesign the autopilot and make it robust, and to introduce a Real-Time Decision (RTD) system to ignite the AS-1, instead of a pre-set time. It is heartening to note that the strap-ons worked successfully on all the four flights. ISRO learned very valuable lessons in launch vehicle technology from the initial failures. The ASLV mission confirmed that failures could indeed serve as stepping stones to success.

The ASLV was thus configured with all solid stages by adding one more stage in the boost phase, to increase the payload capability in LEO from 40 kg to 150 kg. A simpler configuration by strapping two identical motors with core S-1 similar to the core S-1 motor was configured to avoid a fresh development of new motors, which would have called for additional development effort, facility and time. The main configuration change was the canted nozzle (9° canting) to orient the thrust vectors of the strap-on motors pass near to the centre of gravity of the vehicle. I recall the dedicated developmental effort taken up in the PSN and later the SMP, along with all System Development Agencies (SDAs) and processing facilities for the following new technologies/improvements:

- Strap-on joint and jettisoning mechanism.
- Canted nozzle (9° canting) with suitable change in the thermal protection liners.
- Improving the solid loading of propellant to 86 per cent with 18 per cent aluminium fuel content (RCN-01 to RCN-18) in the lower stages. Subsequent to two ASLV failures, there was a mission requirement to bring down the dynamic pressure during transonic regime, which necessitated change in the propellant system in the strap-on motors to switch over to the HTPB-based

propellant system, with burn rates tailored to a higher value of 6.3±0.2 mm/s at the Head End Segment (HES) with associated changes in the motor case Thermal Protection System (TPS) and retaining the same burn rate of 5.1±0.2 in the middle and nozzle end segments (MS and NES).

- Identical performance requirement of the strap-on motors demanded propellant casting adopting the batch concept, throat inserts with identical erosion rate, and tighter control on the fabrication and process procedures to limit deviation within the narrow band.
- To provide a closer prediction of motors performance in flight, the burn rate was estimated using Agni motors and ballistic evaluation motors, which were processed from the same propellant slurry. Triple casting was adopted using the batch process concept. Among these three, two motors were identified for flight and a third motor was tested on the ground prior to flight, and its performance was used to predict that of the flight motors.
- The AS-4 motor case length was extended by 100 mm to load additional 60 kg propellant and the motor case material was changed from E-Glass/Epoxy to Kevlar-49/Epoxy of higher strength, thereby improving the propellant mass ratio from 0.849 to 0.889.

Since there was no requirement for any new development in propulsion systems and control systems, I supplied them as per the designs of the SLV-3, with minor modifications to suit the ASLV needs. The RCS thrusters were upgraded to improve performance and reliability. My team actively participated with the ASLV team in the assembly, integration, checkout and launch operations of all four ASLVs. All propulsion systems requirements of ASLVs were met by the propulsion group of the VSSC under my responsibility. My contributions to the developments of ASLVs are summarized below:

(i) As a member of the project management board of the ASLV team, I actively participated in planning, monitoring, review, guidance and approval of the overall development of ASLVs, with special contributions in the areas of propulsion and mechanical engineering.

(ii) As the director of the propulsion group in the VSSC, for every ASLV launch, I supplied two strap-on solid motors, and solid motors for all the four stages of the ASLV. There was no new development in these systems. I also developed and supplied SITVC systems for the strap-ons and the first stage, the bipropellant RCS for the second stage and the monopropellant RCS for the third stage.

(iii) As the chairman of the standing review committee for the solid rocket motor, I carried out the review of all the solid motor stages of ASLVs – from the design stage to the flight clearance of all the four launches.

(iv) To ensure identical performance of the strap-ons, three solid motors used to be produced under identical process conditions. One of the three was static-tested before flight and the other two were used in the flight. In one of the ASLV launches, the one tested on the ground failed. As chairman of the expert committee, I conducted the failure analysis and concluded that the failure of the ground-tested motor was hardware-specific and that the ASLV flight could proceed with the two strap-ons on hand. The recommendation was accepted by the chairman of ISRO, and the performance of the strap-ons in flight was normal.

PSLV: The Third Launch Vehicle of ISRO

The PSLV Configuration

The Polar Satellite Launch Vehicle (PSLV) (Photo No. 57) was the third launch vehicle of ISRO and the first operational launch vehicle that used liquid propulsion stages. The success of SLV-3 and ASLV provided valuable impetus in its development. I recall that the chairman of ISRO constituted a study group under the leadership of Dr D. Srinivasan in 1978 for research on launch vehicle configurations for an ISRO operational launch vehicle with solid and liquid propulsion stages. The vehicle would be used to place a 600 kg class remote sensing satellite in a Sun-Synchronous Orbit (SSO), also called polar orbit. Dr Srinivasan submitted his first report in 1981 and was appointed as the project director of PSLV.

I remember that it was initially strongly advocated that since ISRO had already established facilities and technologies, as demonstrated in the successful launches of SLV-3 and ASLV, PSLV should also have all solid stages. It is true that at that point of time, ISRO had neither developed high-thrust liquid engines nor established large complex test facilities for their development. But the acquisition of the liquid booster technology from SEP under the ISRO–SEP Vikas agreement influenced the decision to introduce a liquid stage similar to the Ariane second stage. Thus, the first PSLV configuration approved by ISRO in 1983 had four stages, the second stage being powered by a Vikas liquid engine with a nominal thrust of 60 tonnes, burning for a nominal 130 seconds, and a nominal propellant loading of 33 tonnes. However, the conflict between the advocates of the solid propulsion system and the liquid propulsion system surfaced and continued.

In addition, the first stage of the PSLV used two identical liquid engines, each with a thrust of 600 kg for roll control with gimballing capability. While the development of various systems of PSLV were in progress, mission studies pointed out the need for thrust cut-off capability in the final stage to achieve the targeted mission. The advocates of solid propulsion suggested an additional liquid propulsion package for velocity augmentation or cut-off functions. At the same time, the advocates of liquid propulsion got an opportunity to bring out the merits of the liquid stage as the terminal stage. They proposed that the 600 kg liquid engine, being developed for the roll control of the first stage of PSLV be uprated and qualified with a larger expansion nozzle. Two such engines with gimballing capability could be configured as the terminal stage of PSLV. After detailed analysis and discussions, the PSLV configuration was modified in 1985 with a liquid upper stage and twin engines to replace the solid stage as the terminal stage. It was not an easy decision.

I recall the serious discussions that my colleagues and I had with Dr Srinivasan, the project director and his PSLV team at the SHAR guest house in Chennai. I also remember the focused, forceful and logical arguments put forth by Y.S. Rajan, the scientific secretary of ISRO, to highlight the need for a liquid terminal stage for PSLV, which could be a multipurpose stage and used in PSLV missions with different capabilities and applications. Thus, the PSLV Project, with two liquid stages, initiated the beginning of the development of liquid propulsion technology in a major way at ISRO. Accordingly, the finalized configuration of PSLV was a 2.8 m diameter core vehicle having six solid strap-ons of proven 1 m diameter class. The first and third stages were with solid motors while the second and fourth stages had earth-storable liquid engines.

During the development of propulsion systems for PSLV, I focused my efforts mainly on liquid propulsion systems. Nevertheless, I provided guidance and support to solid propulsion developments by Rajaram Nagappa, head of the solid propulsion division of my propulsion group.

The First Stage of PSLV

The first stage motor PS1, named S125 booster, provides the main lift-off thrust, along with either two or four strap-ons out of six. It lifts the vehicle to the altitude of 76 km and imparts a velocity (ΔV^{*}) of around 2 km/sec. The S125 was designed with 129 tonnes of HTPB-based propellant loaded in five segments of 2.8 m diameter and 3.4 m length each. The motor case was designed with high strength M250 grade maraging steel, which was developed in collaboration with the Hyderabad-based public sector undertaking MIDHANI. The details of this solid motor are already presented in the unit on the ISRO solid propellant motors.

The successful static test of S125-ST01 was a great achievement as it symbolized the fruition of totally home-

* ΔV refers to the increase in velocity from start to end of burn.

grown technology. I recall that the magnitude of R&D efforts involved was significant in comparison with 1m diameter motors for SLV/ASLV with respect to size (1 m to 2.8 m), development of case material with almost twice the strength, propellant loading (3 tonnes in a segment to 25 tonnes), thrust capability (45t to 360t) and action time (50 to 100 seconds). Subsequently, the motor was optimized with increased propellant loading and designated S139. With a higher performance, it was successfully qualified and used as the first stage of PSLV from C1 onwards. The S139 became the workhorse booster for PSLV and GSLV missions.

The Second Stage of PSLV

As described earlier, the second and fourth stages of this vehicle were powered by liquid propulsion engines. The second stage was powered by the Vikas engine. In order to develop the second stage for PSLV, in addition to the engine we needed a propellant tank, and structures such as the thrust frame, inter-stages and a number of subsystems, including the pressurization system, fill-and-drain system, gimbal control system, and so on. The ISRO team at SEP obtained the technologies for all elements of the stage except the liquid propellant tank. On their return from SEP, the Vikas fabrication team realized all the elements of the Vikas stage except the propellant tank.

ISRO had decided to fabricate the aluminium alloy propellant tanks and structures at the aerospace division of HAL, Bengaluru. Air Liquid of France had the technology and expertise to manufacture such large aluminium alloy tanks for the Ariane programme. I negotiated with Claude Tronchon, its vice-president, and arrived at an agreement, according to which Air Liquid would manufacture and supply four propellant tanks for the initial flights of the PSLV, according to the design of ISRO. Air Liquid would also transfer the technology and tooling to establish facilities at HAL in Bengaluru, to manufacture subsequent liquid propellant tanks for ISRO. This provision was in addition to building a prototype at Air Liquid and training ISRO and HAL personnel in the manufacturing process, inspection and quality assurance, jigs, fixtures and tooling, and so on. I executed an ISRO–Air Liquid agreement in 1985 with Alain Faure, le chef du service espace of Air Liquid. The agreement was successfully completed, and the aerospace division of HAL supplied all liquid propellant requirements for the PSLV flights. Today, the Aerospace Division of HAL has the requisite technology and facilities to manufacture and supply aluminium tanks for the Vikas stage and other liquid propulsion stages.

Before supplying the Vikas stage for flight, development tests were carried out at component and subsystem levels, and then at the engine qualification level, using battleship configuration and a continuous burn of 150 seconds. Finally, the Vikas stage in the final flight configuration was tested for the full duration at the LPTF stage test stand. The development of the second stage of PSLV was thus completed successfully.

The Third Stage of PSLV

The third stage of the PSLV was designed with several firsts and new state-of-the-art technologies to realize the optimized upper stage with higher propellant mass ratio similar to IUS of the US. The motor had 7.3 tonnes of propellant and 2.0 m diameter motor case wound with high strength (130 kgf/mm^2) and light weight (density: 1.4 gm/cc) Kevlar-49/Epoxy motor case with equal length. The details of this solid propellant motor are presented in the chapter on ISRO solid propellant motors. I am glad to report that the systematic development route adopted by the solid motor team had not derailed the programme in spite of two consecutive failures during the static testing of PS3 third and fourth motors (PS3 ST-03 and ST-04). I had the opportunity to chair frequent review meetings to take real-time decisions during development and at every static testing of the PS3 motor.

The Fourth Stage of PSLV

The fourth stage (Photo No. 58) was powered by an upgraded version of two PS1 roll control engines of 600 kgf thrust. It was totally indigenous and adopted designs and technologies different from the Vikas engine. The Vikas engine had a film-cum-ablative-cooled thrust chamber with a nozzle of silica phenolic throat, impinging multiple radial like-on-unlike injectors, and the propellants were fed by a turbopump driven by a gas generator. But the PS4 engine had a regenerative- cooled thrust chamber with a nozzle of area ratio of 5.24 and a radiation-cooled, columbium alloy nozzle extension of area ratio of 60, multiple element axial triplet injector, and pressure-fed propellants.

The development of this engine started in 1983 but the development tests were successfully completed only in 1988 (Photo No. 59). Upgrading of the RCT to PS4 specifications posed several challenges and optimization of different parameters, as seen from the following comparison of both

55. SLV 3, the first experimental launch vehicle of ISRO, with a payload capability of 40 kg in low earth orbit (LEO). Launched in 1980, with solid propellant motors for all four stages

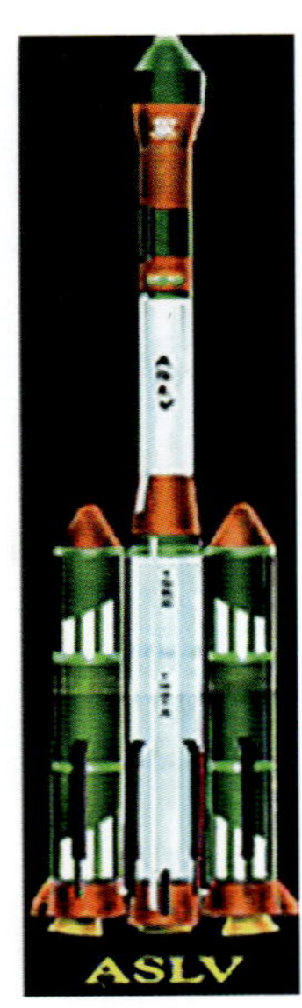

56. ASLV, the second experimental launch vehicle of ISRO with a pay load capability of 150 kg in LEO, successfully launched for the first time in 1992. SLV 3 is augmented with two identical first-stage motors with canted nozzle as strap-on. (Zero-eth stage).

57. PSLV: First operational launch vehicle of ISRO using liquid stages for the first time

58. PSLV fourth stage integrated with control system

59. PSLV fourth stage liquid engine with regeneratively cooled thrust chamber with 45 triplet injectors along with equally spaced 36 orifices along periphery for film cooling and radiation cooled Columbium nozzle divergent with expansion ratio of 60 and mono-methyl hydrazin (MMH) and N_2O_4 propellants

versions. The RCT engine had a thrust of 6.5kN at sea level and operated for less than 110 seconds whereas the thrust of PS4 was 7.5kN at vacuum and operated for more than 420 seconds. The RCT engine was ignited at ground-level and had a combustion efficiency of 90 per cent and Isp of 260 seconds whereas the PS4 engine was ignited at vacuum, had a combustion efficiency of 98 per cent and an Isp of 307 seconds.

Over the years, continuous modifications and improvements were made to optimize the engine performance. The finalized configuration of the PS4 engine had forty-five triplet elements arranged in a zigzag manner over the injector face, along with thirty-six equally spaced orifices along the periphery to provide 30 per cent film cooling. The injector was welded by EBW process to a regenerative-cooled thrust chamber with nozzle of area ratio 5.24. It had a radiation-cooled nozzle extension with area ratio from 5.24 to 60. The injector and thrust chamber were made out of AISI 304, a widely used austenitic chromium-nickel stainless steel, and the nozzle extension from columbium C-103 alloy, which is silica-coated to prevent oxidation at elevated temperatures. The PS4 engine operated at 8.33 bar of chamber pressure with a nominal mixture ratio (MR) of 1.40. However, the same engine, operating as PS1 RCT, had a chamber pressure of 8.0 bar, with the same regenerative-cooled thrust chamber and nozzle with an area ratio of 5.24. Thus, the engine design was the same for both modes of operation.

Since the PS4 engine ignited and operated in vacuum, it had to be qualified for ignition and performance at simulated altitude conditions. The LPTF in Mahendragiri established the test stand for qualifying the engine at sea level and at simulated conditions, and also the stage. To realize a high-performance PS4, we decided to design and develop titanium alloy tanks welded with EBW. We initially designed and manufactured the titanium alloy tanks at our PTU facility in Bengaluru. The PS4 stage – with twin engines, propulsion modules, titanium tank, high pressure gas bottles, RCS thrusters and the integrated PS4 stage in the flight configuration – was successfully tested for full flight duration at the LPTF in 1990.

PSLV Control and Guidance System

The control and guidance system of the vehicle stabilizes it during flight and steers it along the pre-set trajectory profile. For this purpose, the inertial attitude of the vehicle has to be sensed, control command has to be generated to actuate the control power plants, and the control forces and moments required have to be generated to stabilize and steer the vehicle. The responsibility of my team in the control and guidance system was limited to design, develop and supply the control power plants, which would generate the forces and moments. However, we worked closely with the PSLV control and guidance team to ensure the success of every mission.

For the first stage of the PSLV, we supplied the SITVC power plant for the pitch and yaw control. Though this work was based on our experience in the SLV-3/ASLV projects, we had to develop a sophisticated electromechanical valve to handle a flow rate of up to 12 litres per second of strontium perchlorate and meet the closed-loop control requirements. Unlike in the SLV-3/ASLV configuration, we had to distribute these valves in groups of six in each quadrant, totalling twenty-four, and fine-tune them to get the side thrust required. At STEX, we built a six-component static test stand to test the full-scale 125-tonne booster, and we successfully demonstrated the SITVC system. For the roll control of the same first stage, we developed and supplied two liquid engines of 600 kgf thrust each, which were capable of swivelling and meeting the control requirements. Through a series of ground tests, we demonstrated this roll control system with two liquid engines, pressurization system, propellant tanks, regulation system, rotary actuators for swivelling, thermal management system and so on.

My Contribution to PSLV

(i) As a member of the project management board of the PSLV team, I actively participated in the planning, monitoring, review, guidance and approval of the overall development, with special contributions in the areas of propulsion and mechanical engineering.

(ii) As the director of LPPs in the VSSC, I developed and supplied the second stage of PSLV, the first liquid stage introduced in the ISRO launch vehicle programme. Today, the second stage of PSLV has a N_2O_4/UH25 propellant loading of 42 tonnes developing about 80 tonnes thrust in flight and burns for about 150 seconds.

(iii) As the director of the propulsion group in the VSSC, I developed and supplied the fourth stage of PSLV with twin engine configuration each developing about 7.35kN in flight, using MON/MMH as propellants with a nominal propellant loading of 2 tonnes. In addition to the main propulsion, the stage had six

5 kgf thrusters, which used the same propellants for RCS. This stage, being the terminal stage of the PSLV, was upgraded with increased propellant loading and burnt for 333 to 526 seconds as required in flight.

(iv) As the programme director of APSU, I was responsible for the development and supply of:

  (a) the SITVC system for the first stage of PSLV,
  (b) the SITVC system in strap-on stages (PS0) for roll augmentation, and
  (c) RCS for the PSLV first stage.

As the programme director of APSU, I also supplied a large number of components and modules for the liquid stages and reaction control systems. For a single PSLV flight, APSU supplied 243 components and modules.

(v) My group members joined the PSLV team and contributed in the stage assembly and integration activities in the PSLV assembly hall and also in the launch complex. In addition, we participated in the launch campaign for the propellant servicing of both the second and fourth stages of PSLV and also the PSLV control systems.

PSLV Flights

ISRO planned three developmental flights before making the PSLV operational. The developmental phase of around twelve years was complex and difficult. After the successful completion of all the developments, the first developmental flight PSLV-D1 was launched in October 1993 but it could not meet the mission objectives. The failure was analysed and remedial actions were taken. The next developmental flight PSLV-D2, launched in October 1994, was successful as it met all the mission objectives. This feat signalled ISRO's entry into the operational launch vehicle league. The PSLV-D3 launch in 1995 successfully completed the developmental phase.

PSLV commenced its operational phase with the successful launch of PSLV-C1 in September 1997. With a large number of successful flights and the launch of Indian and foreign satellites, PSLV is now recognized globally as a reliable satellite launch vehicle and is the pride of ISRO.

15

Out of the ISRO Orbit: The Pain of Ignominy

I WAS UNDOUBTEDLY THE senior-most centre director at ISRO, with around twenty-nine years of meritorious service in the organization. Nevertheless, in April 1994, I was bypassed and Dr K. Kasturirangan was appointed as secretary to DOS, and the chairman of the Space Commission and of ISRO. I felt that I had lost a prestigious race and decided to quit ISRO.

However, on hearing the news about Dr Kasturirangan being appointed as Dr U.R. Rao's successor, I called and congratulated him. I also contacted Dr U.R. Rao, thanked him for his guidance and requested a month's (well-earned) leave to plan my future. It was granted. Though I was not officially on duty during this period, I attended to crucial issues related to the ASLV launch scheduled for May 1994, the PSLV launch planned for October 1994 and the subsystems under preparation for IRS-1C, INSAT-IIC and PSLV-D3.

When all of this took place, I was not sure what the future held, but I believed that the Almighty had a different plan for me. The subsequent assignment with the Government of India that desired to retain my services to the nation came as a welcome relief.

However, I wish to place on record the course of events that led to my decision to quit. The denial of timely and eligible promotions, which adversely affected my career growth, the obvious failure to recommend me for national recognition and what seemed to me as favouritism by the chairman – all had a role to play.

My Role Models

Dr Arcot Ramachandran and Dr Vikram Sarabhai are well-known personalities and reputed sons of our nation. As mentioned previously, they have been my role models. The former inspired me to leave the country for higher studies, and the latter was instrumental in my returning to build an enriching and rewarding space programme for India. But they did more than this; I would like to briefly reveal the nurturing influence that these two eminent great men had on my career and personality.

Dr Arcot Ramachandran motivated me to take up higher learning and instilled in me the thirst for application-oriented research and development. At the age of forty-four, he became the second director of IIT Madras and contributed significantly to the development of the institution in its nascent stages. In 1973, he was invited by the Government of India to be the first person to assume the post of secretary in the Department of Science and Technology. In 1998, the United Nations appointed him the executive director of the

United Nations Centre for Human Settlements, at the level of an under-secretary general. On his retirement, he settled in Bengaluru. I maintained contact with him till his death.

Prominent among the many values he imparted was the importance of deep subject knowledge, which is always relevant at work. He taught me to analyse all options, seek expert guidance and consult colleagues before arriving at significant resolutions. I learnt to take firm and quick decisions, to communicate them with clarity to those concerned and to follow up on these matters regularly. To monitor, review and guide the progress of each venture and initiate corrective actions where necessary became second nature to me. I have tried to be just in every action and cordial in my interactions with others.

As readers may remember, Dr Vikram A. Sarabhai was my first boss in India. My career at INCOSPAR/ISRO began in 1966 as the divisional head of propulsion engineering and mechanical engineering at the SSTC at Thumba, of which Dr Sarabhai was the founder director. I worked very closely with him as a subordinate until his unexpected demise. I have already presented lasting memories of my association with him over the brief period that I spent with him. The five years of knowing him seemed like a lifelong relationship to the enthusiastic youth in me. Dr Sarabhai adopted a benevolent approach to all problems and respected everyone as human beings. From him, I imbibed the art of building a team with competent people to accomplish tasks in the spirit of cooperation and to value the contribution of every member even though all are not equally competent.

I followed his pragmatic approach to evolve action plans to achieve set targets based on the understanding of available resources, including infrastructure, human resources, finance and limitations, if any. Consequently, I have strived to carry team members along with me so as to function as a family together, often involving them in taking collective decisions and to be transparent in all actions. Dr Sarabhai had the rare skill of making every employee appreciate the facts and circumstances of a situation – thus getting things done through requests and not by issuing orders as a senior officer.

My Modus Operandi

As the first member of the propulsion system development team in ISRO, I established the relevant activities over time. The major task then was building human resources. I took personal interest in it and for many years I knew the names of every single member on my team of about 600. Initially, I used to visit the laboratories and workshops daily and say a few words to the employees, individually as well as collectively. I maintained some sort of personal relationship with them, which enhanced the morale of the workforce.

No one needed prior appointment to meet me in my office. Any individual could check with my personal assistant and walk in unless I was otherwise engaged. I had a good understanding of all that occurred on the premises. When someone raised a problem, I was often aware of the details beforehand. If not, I would call the people concerned to my office and discuss the issue without delay. In most cases, we arrived at a satisfactory solution. If necessary, I would follow this up with an office order. When a larger consultation was required, I would take it up in our weekly meeting at the centre.

Under my chairmanship, I used to conduct a weekly centre meeting at the VSSC every Monday afternoon with all the divisional heads, senior engineers and heads of administrative services. The head of Personnel and General Administration (PGA) was the secretary. He used to prepare the agenda for the meeting and maintain the records. If any Monday happened to be a holiday, the meeting would be held on the next working day. We used this opportunity to review the status and progress of all technical and administrative matters and to finalize our plans for the week ahead. Everyone was allowed to voice their opinions and to bring in new proposals. The meeting would last until all the items on the agenda were covered – another habit I picked up from Dr Sarabhai. At times, the discussions would extend late into the evening, and tea and snacks would be served. This system was appreciated by everyone at the centre; it considerably helped our technical activities and also redressed any administrative problems.

I would like to mention an interesting interaction that occurred at one of those gatherings. At times, the issues discussed were complex and everyone would not be satisfied with the outcome. During one of the meetings, when I conveyed my decision and identified the follow-up action plan, one of my senior divisional heads asked me: 'Sir, is it a request or an order?' Apparently, he had not found my verdict favourable. I replied with a smile: 'Gentleman, it is a request, if complied with. If not, an order.' Everyone smiled, the ice melted, and my opinion was accepted without the need for an office order.

My style of operation was obviously different from many operational government systems of the time. Performance held priority over procedure, without any violation of rules. Many appreciated my method of functioning but some had reservations. They felt that the boss should be a bit more reserved and maintain some distance from colleagues. He should not be too transparent and disclose all the facts. Undaunted, I continued with my conviction, functioned in a fair manner and implemented the strategies I had picked up from my role models.

An Exemplary Career

When Dr Sarabhai appointed me, a Scientist/Engineer D (SD) grade employee, as the head of both propulsion engineering and mechanical engineering divisions with effect from 15 April 1966, it was unique in itself. No other head at the SSTC then was in charge of two divisions, even though all of them enjoyed higher grades. A few months after I joined, Dr Sarabhai informed me that I would be normalized in my grade after a year of service. Accordingly, along with H.G.S. Murthy, the director of TERLS, he met me after I completed one year. They assessed my performance and promoted me to the Scientist/Engineer E (SE) grade from 15 April 1967.

ISRO had a three-year assessment and promotion policy then. Accordingly, Dr Sarabhai promoted me to Scientist/Engineer F (SF) grade with effect from 15 April 1969, the very day I completed three years of service. Subsequently, there was an overall review of the ISRO promotion policy and promotions were effected for all seniors who had completed two years of meritorious service. As per the revised policy, my senior colleagues at the Scientist/Engineer F (SF) grade were promoted to Scientist/Engineer G (SG) but I was promoted, in a deviation from normal procedure, from SF to Scientist/Engineer F grade with effect from 15 April 1971. I was the only senior scientific person in the F grade then at ISRO and continued to be so until the next promotion. Thus, history was again created in my career at ISRO for reasons not known. Only later did I realize that this uniqueness of my position could go against my career growth.

When I discussed this anomaly with Dr Sarabhai, he explained to me that there existed two promotion avenues – the first one was SE-SF-SG-G-H and so on, and the second was SE-SF-F-G-H and so on. Dr Sarabhai told me that I was being taken along the second avenue for reasons that he did not disclose. I accepted his words and continued with my work. Apparently, it was a 'management strategy', since a few who had started work at higher grades opposed the rise in my career. They recognized that I had more challenges and opportunities as the head of two divisions.

Following Dr Sarabhai's unexpected demise, there was a change in the top levels of ISRO management. Modifications were made in the promotion policy by introducing 1 January and 1 July as the dates for review and promotions every year, irrespective of an employee's joining date. Accordingly, a few promotions were accorded to senior personnel in 1973. I was thus promoted to the SG grade, instead of the higher G grade that I was originally due to be promoted to, in January that year. This move was against Dr Sarabhai's ways.

Further, the joint secretary of the DOS issued a pay revision order that revamped the pay scales in ISRO, based on the Third Pay Commission orders.* According to the order, grade F was abolished from 1 January 1973 and merged with the SG grade. Thus, the ISRO promotion order that had raised me from F grade to SG grade had no validity since the two grades were merged on the date of the promotion order.

---

* Vide Order No. G/11013-2/ISRO-74 dated 6 July 1974.

Essentially, I was denied a promotion, and I was the only one adversely affected in the organization. Unfortunately, Dr Sarabhai was not around to implement his promise.

Since these developments had taken place due to administrative procedural delays in pay revision orders, it could have been sorted out in the department. So I brought this anomaly to the attention of Dr Dhawan, the chairman of ISRO. But there was no response. Thus, for no fault of mine, the department denied me an eligible promotion, which was available to all my colleagues.

In spite of this unfair and disappointing turn of events, I continued to contribute to the programmes of ISRO in the best possible manner. But I couldn't do so for long. In course of time, on 1 January 1980, many of my junior colleagues advanced to the G grade. This opportunity too was denied to me. I was very upset, not because I was refused the promotion but because I had started to lose respect in the eyes of my juniors. I took leave and planned to explore other options outside ISRO. I even considered shifting to academics, which was once my favourite profession.

This move of mine came to the notice of Dr Dhawan. During his next visit to the VSSC, he asked me to meet him at the Thiruvananthapuram airport as he waited for the flight to Bengaluru. I had been on casual leave, but I met him as instructed. He wanted to talk to me alone. We walked along the seashore. I told him: 'Sir, you are the boss. You make decisions in the interest of ISRO. If I cannot accept your decision, my only option is to quit. I have nothing against you or the system. In my opinion, injustice was done to me.' He patiently listened to me and agreed to look into the issue and set matters right. He appeared to have realized that there was some truth in my words. Later, I was told that he had asked the director of the VSSC for the details of my case. He even processed the proposal for my promotion and placed me in the G grade with effect from 26 April 1980, as a special case. The denial of an eligible promotion due on 1 January 1980 was a sign that the ISRO management may not be fair in extending my prospects in the organization.

I got my subsequent promotion to H grade on 1 January 1985. Years passed by, and further advancement was delayed again for a while.

All that would change for some time, though. I heard that the chairman of ISRO had proposed to the government that all directors of the major ISRO centres should be on par with regard to their salary and grade. Accordingly, all the centre directors were to be granted the Outstanding Scientist grade with effect from 1 January 1992. However, the Appointments Committee of the Cabinet approved the proposal from the chairman with a modification – that Dr A.E. Muthunayagam be promoted six months ahead of the others. This step made me senior to all the directors of the ISRO centres and instilled in me a false hope for justice in my career.

I was aware that promotion is not a birthright, or something that all employees are entitled to, and that no one could claim exceptional opportunities especially as a right. It is primarily policy-based and a decision of the management concerned. Nevertheless, it is natural for employees to expect equal treatment from the institutions they work for. From the experiences so far, I had reached the conclusion that the ISRO management would deny the kind of prospects that I deserved and hinder my career, which ought to have culminated in the chairmanship of the organization.

Denial of National Recognitions

My contributions to ISRO and its space programmes were also not recognized at the national level on time. I understood much later that such accolades depended upon factors other than merit and individual performance. I realized that it is not what you are that counts, but it is what others think what you are that matters. Based on the recommendations of the ISRO management, the Padma awards were announced by the Government of India in connection with the Republic Day celebrations every year. Prior to the PSLV launch, Dr U.R. Rao told me that I had contributed significantly to the development of the second and fourth liquid stages of PSLV, in addition to the work done on solid propulsion systems and control systems. He added that ISRO recognized and valued my efforts very much. He gave me the impression that the organization would nominate me for the Padma awards after the PSLV launch. But I have not been awarded any Padma to date. I am convinced that my contribution to the Indian Space Programme has not been less than that of many other ISRO personnel who were given the Padma awards.

Dr U.R. Rao, who took over as secretary to the DOS, the chairman of the Space Commission and of ISRO in 1984 from Dr Dhawan, was my colleague at the SSTC under Dr Sarabhai. In ISRO's infancy, Dr U.R. Rao and I worked under Dr Sarabhai as divisional heads; he headed the Satellite System Division (SSD) and I the Propulsion Engineering Division (PSN) and Mechanical Engineering Division (MEC). Our offices were on the same floor as that of Dr Sarabhai.

Subsequently, Dr U.R. Rao moved to Bengaluru as the project director of the ISRO Satellite System Project (ISSP).

Over a period of time, the SSD at the SSTC was wound up and all satellite-related activities were shifted to Bengaluru. Dr Rao exploited the challenges and opportunities in the nascent satellite field with excellent guidance and support from Dr Dhawan, the new chairman. He went on to establish the ISRO Satellite Centre (ISAC) on 3 November 1976 and became its founder director. Eventually, he rose through the ranks and became the chairman of ISRO.

Aryabhata was the first satellite project of ISRO, and Dr U.R. Rao was the project director. Dr Kasturirangan joined ISRO Satellite System Project (ISSP) and assisted Dr U.R. Rao in the project. In the early days, ISRO had been developing satellites for both remote sensing (polar) and communication (geosynchronous) applications. After the launch of Aryabhata on 19 April 1975, Dr U.R. Rao oversaw the launch of the second remote sensing satellite Bhaskara, with Dr Kasturirangan as the project director, and the Experimental Communication Project APPLE, with R. Manickavasagam as project director.

Succession Planning by Chairman

When Dr U.R. Rao became the chairman of ISRO in 1984, Colonel N. Pant became the director of ISAC. This was when a series of transfers took place. First, officials who were senior to Dr Kasturirangan were shifted out of ISAC in a phased manner until he was made its director. For instance, in 1985, Manickavasagam, the project director of APPLE, successfully completed the geosynchronous experimental project and was shifted from ISAC to the VSSC.

In 1986, Dr S.P. Kosta another engineer who was also senior to Dr K. Kasturirangan and had contributed to satellite projects in the initial stages was shifted out. Thus, Dr Kasturirangan became the senior-most officer in ISAC. In 1986, he was made the associate director of ISAC and his former responsibility of overseeing the operational remote sensing satellite IRS was given to S. Kalyanaraman of ISAC.

Finally, in 1990, when the post of the vice chairman of ISRO was created to assist the ISRO chairman, the then director of ISAC, Colonel N. Pant, was given the role and shifted to the headquarters. Dr Kasturirangan then became the director of ISAC in his place, in 1990. It is noteworthy that in 1991, Dr U.R. Rao abolished the post of vice-chairman when Col. Pant retired from government service on superannuation.

As is obvious from this unnatural and continuous movement of officials to various positions within the organization, officials who were senior to Dr Kasturirangan were shifted out of ISAC in a phased manner until he was made its director. My eligible promotions were delayed for reasons not known. However, Dr Rao promoted Dr Kasturirangan by giving accelerated merit promotion twice in the space of three-and-a-half years from grade G to H in July 1988, and another merit promotion in the following three-and-a-half years from grade H to Outstanding Scientist in January 1992.

This decision came when Dr U.R. Rao had proposed to the government to make all major centre directors on a par regarding their grade and salary. It was a veiled attempt to make Dr Kasturirangan equal to me in grade and pay. However, the Appointments Committee's amendment – that I should be at a level six months ahead of all centre directors because of my seniority, contributions and long years of service at top positions – stymied Dr Rao's apparent manipulations. The entire fiasco is exposed by the fact that none other than Dr Kasturirangan has received two successive merit promotions after three-and-a-half years of service in each grade in the history of ISRO!

In April 1993, the Space Commission for 1993 was formed with increased membership from seven to nine, another astute strategy to elevate Dr Kasturirangan above others. Dr S. C. Gupta, Director, VSSC, the senior-most director at ISRO who was due to superannuate in January 1994, and Dr Kasturirangan, Director, ISAC, junior-most director at ISRO, were added to justify the inclusion of representation from launch vehicle and satellite divisions. Dr Gupta had merely nine more months of service to superannuate, whereas I had more than six years left at ISRO then. Further, the Space Commission for 1994 was reconstituted in January 1994 itself with the same membership as that of 1993. Dr Gupta had only one more month of service at ISRO at that juncture. As a result, I was denied the opportunity to become a member of the Space Commission.

The VSSC is the largest and oldest centre of ISRO and is also the lead facility for launch vehicles. LPSC contributes in a major way to launch vehicle developments. It is the norm that the director, LPSC, becomes the director, VSSC, when a vacancy arises – the appointment of G. Madhavan Nair, S. Ramakrishnan, M.C. Dattan, Dr K. Sivan and S. Somnath is proof of this convention. When Dr Gupta, Director, VSSC, superannuated in January 1994, P.P. Kale, Director, Space Application Centre (SAC), became the director, VSSC, thereby rejecting my written request to Dr U.R. Rao for a posting at the VSSC. It must be reiterated that I was the senior-most ISRO director, with nearly twenty-eight years of experience in launch vehicle developments, whereas P.P. Kale was junior to me and had expertise in communication, electronics and satellite applications.

Prof. Dhawan discussed my future at ISRO with me. He told me that ISRO had a major task ahead in the area of launch vehicles, particularly in the cryogenic stage development that I had initiated. He suggested that I focus on this area and take total responsibility for launch vehicle development and assist the chairman, ISRO. I was shocked; nevertheless, I respected his advice and told him that my ability to perform a task creditably should not stand in the way of my career growth. I felt that I should not be deprived of what I deserved. I would continue to work at ISRO if the successor to Dr U.R. Rao was more capable and better suited for the position of chairman than me. If I was not convinced, I would have to quit my job.

The issue of a successor to Dr U.R. Rao became a hot topic of discussion at ISRO. I was certainly deemed one of the eligible candidates. The four major ISRO centre directors in the order of seniority were: Dr Muthunayagam, Director, LPSC, P.P. Kale, Director, VSSC, R. Aravamudan, Director, SHAR and Dr Kasturirangan, Director, ISAC.

During the discussions with Dr U.R. Rao, P.P. Kale told him that he would like to continue with ISRO if Dr A.E. Muthunayagam, the senior-most candidate, was appointed as the successor to Dr Rao. If not, he was ready to leave ISRO. True to his word, P.P. Kale took voluntary retirement when Dr Kasturirangan was eventually appointed as chairman, ISRO.

As things turned out, Dr U.R. Rao refused me the coveted position that would have been the crowning glory of my career.

Auxiliary Propulsion System Unit: An Oasis

As we progressed with major achievements in all propulsion system development activities being my responsibility, there were avoidable delays in not integrating the work under an umbrella and refusing autonomy and authority to the leadership. This state of affairs stood in the way of an efficient mode of functioning. It appeared that the director of the VSSC, was against parting with any development responsibility on liquid system developments though all such work was carried out under my supervision in the various projects of the VSSC. Apparently, he wanted to control and regulate all my activities, but I had commitments to supply pressure transducers to SEP under the Vikas agreement with a penalty for delays. The ISRO management assessed the existing work relationships and evolved a strategy that ensured partial autonomy to discharge my duties.

In 1980, when the liquid propulsion project was established at the VSSC with me as the project director, another organizational change was made to create the Auxiliary Propulsion System Unit (APSU), also under my responsibility, as an autonomous unit of ISRO. The management intended to group together some of the activities related to the development of propulsion system that I had been executing at the VSSC and PTU. I was designated the programme director of APSU, delegated the powers of a head of the department at DOS and asked to report to the chairman of ISRO. This position gave me a certain degree of autonomy in the operations within DOS and allowed me to take decisions to accelerate the developments at APSU.

As we advanced with laudable achievements in propulsion system development, all the activities at APSU and the VSSC were integrated under the Liquid Propulsion System Unit (LPSU) in 1985, of which I became the director with full autonomy in operations. Further, on 1 June 1987, the LPSU was renamed the Liquid Propulsion System Centre (LPSC) without any change in the management system. Such a two-step elevation in the organizational status was the first instance of its kind in the history of ISRO. To sum up, my responsibility in propulsion system development began in 1966 at the SSTC as its divisional head and continued at the VSSC from 1972 as the group director, propulsion, project director, Vikas, and project director, liquid propulsion projects, with all these positions reporting to the director of the VSSC. In 1980, I became the programme director of APSU and reported to the chairman ISRO, and ultimately instituted an R&D centre for liquid propulsion systems in 1987. Indeed, a long and remarkable journey.

Exploring Options

In April 1994, I discussed my future options with Dr Kasturirangan, the new chairman of ISRO. We agreed that I ought to seek an alternative assignment with the Government of India. Accordingly, on 18 April 1994, I appealed through him to Prime Minister Narasimha Rao to consider me for a suitable assignment in the scientific departments of the government, which included the departments of Defence R&D, Science and Technology, Ocean Development and Non-Conventional Energy Sources. I also presented my case to Bhuvanesh Chaturvedi, the minister of state for space, during his visit to the VSSC. Dr Kasturirangan told me that he would take up my case with other government departments too. I had expressed my preference for an assignment in an executive position and not in an advisory capacity – as 'officer on special duty assisting VIPs' or 'member of national-level commissions', to name a couple of such posts.

The Controversial Espionage Case and the CBI Investigation

In October 1994, a false allegation of 'espionage' was made against the LPSC and two senior officials – S. Nambinarayanan, project director of the Cryogenic System Project (CSP) and D. Sasikumar, deputy project director of engine fabrication in CSP – were suspended from service by means of a direct order of the controller, LPSC. I was the director of the LPSC and chairman of the management board of CSP. The two officials mentioned here were my subordinates. However, I was not informed of any details of the case or officially consulted in this regard.

The Kerala Police conducted the initial investigations; subsequently, the CBI took over the case. The enquiry went on at different levels. Finally, the Supreme Court of India acquitted both Sasikumar and Nambinarayanan in 1998. On Nambinarayanan's petition seeking compensation, the apex court reiterated that he had been wrongly accused in the matter and directed the Government of Kerala to pay him Rs 50 lakh as compensation.*

Regrettably, in 1994, when criminal charges were framed against the LPSC, the joint secretary of DOS, who was also

---

* Judgment of the three-judge bench of the Supreme Court presided over by CJI Deepak Mishra on 14 September 2018 *in S. Nambi Narayanan, Appellant(s) VERSUS Siby Mathews & Others Etc.,Respondent(s)*, CIVIL APPEAL Nos. 6637-6638 of 2018.

its chief vigilance officer, bypassed me, the director of the centre, and dealt directly with the Kerala Police in association with the controller of the LPSC. Dr Kasturirangan, who had taken over his assignment as secretary in DOS about six months earlier, was new to the LPSC. He was not familiar with the operational system and the procedures in place there. He appeared to lack confidence in me because I was once his senior and a former unwelcome competitor to his new position. This is probably why he did not discuss the case with me.

From the information available in the public domain, it appeared that this case came under the Official Secrets Act, which was within the jurisdiction of the Department of Space, the Government of India (GOI), and not the Kerala Police. The entire affair could have been solved in the initial stages if Dr Kasturirangan had exercised his authority as the secretary to DOS, taken the director of the LPSC into confidence, investigated the case within the department, ascertained facts and acted in a just and responsible manner. The police interrogation, subsequent torture and suffering of two senior officials of ISRO, both of whom had rendered significant contributions to its prestigious programmes, could have been avoided – apart from the irreparable loss of their dignity and reputation in society.

Report to the Chairman of ISRO from LPSC

For about two months since August 1994, I carried out a detailed stock verification of all technical documents and drawings at the LPSC. I found that all the original and basic technical documents and drawings related to liquid propulsion system developments and the day-to-day functioning of the LPSC were in stock. These documents and drawings were associated with design, analysis, fabrication process/drawings, reliability and quality control, assembly and integration, test and evaluation, and so on. There was no loss of documents/drawings to adversely affect the plans and programmes at hand then. I brought this fact to the attention of Dr Kasturirangan. I further informed him that the developments of the liquid propulsion system were carried out in different groups/divisions/facilities of the LPSC, depending on the expertise and requirements.

The documentation system was decentralized and the responsibility for generation, storage and issue were delegated to the group/division in the LPSC. It was observed then that no ISRO centre had in operation a system and procedure to classify the technical documents/drawings. Moreover, no specific procedure was stipulated at ISRO on the duplication, issue or control of documents. LPSC followed the system and procedure in operation in other ISRO centres. Further, the plans and programmes of ISRO were well publicized and available in the public domain. There was no secrecy involved. The R&D centres of ISRO interacted with academic institutions and industries and utilized their resources. Transparency in technical interaction with external agencies and the free exchange of documents accelerated the progress of activities.

In a nutshell, there was no control on the exchange of documents. I also reminded Dr Kasturirangan about the meeting he had with centre directors, unit heads and other senior officials of ISRO wherein the urgency to formulate organizational-level policies and guidelines for the classification of technical documents and drawings, and the procedure for the control, handling, duplication and distribution, periodic verification and reclassification of technical documents and drawings were also discussed. This matter was taken up by the ISRO headquarters. However, during my tenure as the director of the LPSC, I did not receive any orders on the topic from ISRO authorities. I left ISRO on 14 April 1995.

The Decision to Quit ISRO

Dr Kasturirangan did not consult me on any of the LPSC programmes, particularly the cryogenic stage development for the GSLV. He had his own strategy; he suggested that I be shifted to Bengaluru as his adviser, with the specific assignment and role to be defined later. I considered it an insult and a humiliation, as any change in my position could only come with greater responsibility and accountability. I needed to think about his suggestion. In January 1995, I informed him of my preference to quit ISRO instead and requested a month's time to finalize my plans.

Exploring Overseas Assignments

The response from the government to my request for an alternative position was delayed. Meanwhile, I explored the possibility of working outside the government and looked up options for short-term assignments abroad. Consequently, in May 1994, with the approval of DOS, I approached the National Council for Scientific and Technological Development, (the Conselho Nacional de Desenvolvimento Científico e Tecnológico [CNPq]), Brazil.

In the meantime, I learnt from a reliable source in Delhi – at the level of additional secretary to the Government of India

– that the Department of Science and Technology (DST) was asked to keep a good room in their guest house ready for Dr Muthunayagam from ISRO to be posted as a secretary to the Government of India. Accordingly, DST made suitable arrangements at their RK Puram guest house. Unfortunately, it was around that time that the ISRO espionage case appeared in the newspapers, and consequently the appointment was kept in abeyance. I approached A.N. Varma, Principal Secretary to the Prime Minister, on 21 January 1995 and explained my predicament. He told me that the Government of India required my services and that my posting was under active consideration.

'Visiting Foreign Scientist'

In the first week of February 1995, I received the offer of a scholarship as a visiting foreign scientist for a period of one year from 1 February, which could be extended for another year. The scholarship came with an annual remuneration of USD 30,000, along with free accommodation, medical insurance and travel expenditure. The assignment was at the Brazilian General Command for Aerospace Technology (CTA). I had two months to arrange for my visa and travel. I informed the chairman of ISRO about the offer.

On 14 February, I requested official approval from ISRO to take up this short-term assignment in Brazil on deputation, initially for a period of one year and extendable for another term. To avoid procedural delay, on the next day, i.e., 15 February 1994, I requested for extraordinary leave for the same period instead of the deputation. I received no response from the chairman for nearly a month. Since

I had to communicate with CNPq without further delay, I approached A.N. Varma once more on 14 March 1995 with a request to either give me an alternative assignment in any of the government departments or grant EOL before the end of the month.

Secretary to the Government of India

On 5 April 1995, the Appointments Committee of the Cabinet conveyed the approval of my appointment as secretary in the Department of Ocean Development (DOD) with effect from the date I assumed charge. On 14 April 1995, I had a brief meeting with G. Madhavan Nair (the LPSC director designate), M.K.C. Nair (controller, LPSC) and my office staff. I handed over my duties to G. Madhavan Nair and was relieved on the same day. On the last working day of my life at ISRO, I recalled all the memorable moments of the past three decades and placed on record my sincere thanks to all my colleagues, friends and well-wishers who guided, supported and stood by me during both pleasant and difficult times at ISRO.

On 14 April 1995, returning home as an ordinary citizen with no assignment in the government, I thought about my family that had supported me during my long tenure at ISRO. As a husband and father of three sons, I felt that I did not do full justice to my family. I was the only breadwinner. My wife took full responsibility in managing the day-to-day functioning of the household and supervised the education of our children. While I admitted my children in the best schools in Thiruvananthapuram, it was my wife who looked after them on a regular basis. I rarely attended the school functions of my kids. On Sundays and holidays, we visited friends in the afternoons and spent some time enjoying the social interaction.

About a dozen families from Nagercoil that had settled in Thiruvananthapuram used to have get-togethers once a month and on special occasions such as birthdays and anniversaries. Then all the families met in one house and had a good time until we dispersed after a wholesome dinner. This pattern slowly faded into memory as some of the elders passed away and the children grew up, went to college and later, took up jobs in distant places. In spite of my official commitments, we participated in most functions, including the weddings in my native place. I attended church with the family every Sunday and celebrated Christmas, New Year's Day, Easter and so on. Looking back, I am filled with gratitude to my family members for their unstinted support and prayers.

I took charge as the secretary, Department of Ocean Development (DOD), Government of India, on 17 April 1995. The first day of my new assignment in New Delhi turned out to be memorable on two counts. The first was chairing the Planning Committee meeting for the Annual Antarctic Expedition in 1995. The meeting was organized by J.V.R. Prasad Rao, joint secretary of DOD. He meticulously charted out every aspect of the expedition, gave me the opportunity to know more about the discipline and let me meet the directors and senior officials from different R&D centres who participated in the national Antarctic programmes. It was a new and thrilling experience.

The second event was paying my respects to the Prime Minister of India who was also the minister in charge of the department. Prasad Rao arranged the appointment. I was chairing the Antarctic meeting when the call came from the Prime Minister's Office. I excused myself and drove to South Block. I was meeting Prime Minister Narasimha Rao for the first time. When I presented him with a bouquet and paid my respects, he said, 'We will meet and discuss your programmes after you settle down.' He added, 'We need competent people to work for the government in Delhi.' I was amazed. It appeared that the Prime Minister was aware of the hurdles I had crossed before arriving at the capital to assist in the ocean programmes of our nation.

From 'Space' to 'Ocean'

Propulsion systems, after completing their mission in space, end up in the ocean. Ironically enough, I got into a similar scenario! Nevertheless, unlike the propulsion systems which sink to the bottom of the sea, I swam along with others and contributed to national and international ocean programmes in accordance with the Ocean Policy of the Government of India. My performance and input as a sailor in the tumultuous waves of the Department of Ocean Development are presented in the following chapters.

16

Orbit Transfer: From Space to Ocean

OPTING OUT OF THE 'space' orbit, I regulated my 'career opportunity velocity' and entered the 'ocean' on 17 April 1995, when I was assigned the responsibility of heading the Department of Ocean Development (DOD) as secretary to the Government of India. What a transition – from propulsion systems for spacecrafts and launch vehicles to ocean resources! DOD was created in July 1981, as part of the Cabinet Secretariat, directly under the charge of the Prime Minister of India. From March 1982, it became a separate department of the Government of India.

During my tenure (April 1995–January 2001), I had the opportunity to work under four Prime Ministers – P.V. Narasimha Rao, I.K. Gujral, H.D. Deve Gowda and A.B. Vajpayee – and subsequently under Dr Murli Manohar Joshi, the minister for ocean development, science and technology, and human resources development. All these leaders provided the impetus and support necessary to make the DOD a pioneer in advanced technologies to use ocean resources.

The oceans are known to be our last frontier on the planet. The programmes of the department were formulated based on the 1981 Ocean Policy Statement of the Government of India and the Allocation of Business Rules 1961, amended from time to time, and activities specifically assigned by the government. The responsibilities of the department were to formulate and implement the programmes with long-term economic and technological implications, and to act as a nodal agency for technology development in areas relevant to the economic and social development of the coastal population. It was also meant to act as a body to coordinate scientific and technological programmes for the exploration of living and non-living resources of the ocean and the preservation of the marine environment.

The implementation of the United Nations Convention on the Law of the Sea (UNCLOS), which came into force in 1994, provided an extended area of national jurisdiction in the seas and an increased base for the exploration of living and non-living resources in the Exclusive Economic Zone (EEZ) and the continental shelf. The United Nations Conference on Environment and Development (UNCED), 1992, suggested an integrated approach towards environment and development, which included coastal zone development and management. Consequently, during my tenure, the department was augmented and its activities and programmes were reoriented towards the exploration and sustainable exploitation of marine living and non-living resources. The protection of the marine environment with the active participation of national R&D

laboratories, educational institutions and industries was an innovative venture.

A serious limitation that the department faced during the early 1990s in the implementation of priority programmes was the lack of institutional infrastructure directly under its management. The Ocean Policy Statement envisaged that the complex programmes of ocean development would entail a well-designated management and institutional extension of DOD, with sufficient powers vis-à-vis other agencies to ensure speedy developments in the ocean sector. India was to be at the forefront of the international effort. Moreover, the coordinating role of the department prior to the Eighth Five-Year Plan in relation to other ministries and departments could not be pursued effectively in the absence of an apex body such as the ocean science and technology agency/board, the ocean commission, or autonomous institutions under the department.

Eighth Five-Year Plan Initiatives (1992–97)

During the Eighth Five-Year Plan period (1992–97), the department envisaged the implementation of programmes in line with the Ocean Policy Statement and to strengthen the institutional infrastructure. The department had three wings, all reporting to me – administration with a joint secretary (JS) (Administration), finance with a joint secretary (JS) (Finance), and science and technology, with a few scientists and technologists who coordinated and executed multi-institutional and multidisciplinary projects in ocean science and technology, in association with academic institutions, national laboratories and ministries.

The DOD operated on a small budget for the Antarctic research and expedition, oceanographic research and capacity building. When I joined the department, the major activity was the Antarctic expedition and associated research, for which about 70 per cent of the budget was allocated. Nearly 15 per cent was sanctioned for oceanographic research at the National Institute of Oceanography (NIO), CSIR, Goa. The remaining 15 per cent was meant for expenditure on administration and international programmes. I felt the need to strengthen the science and technology focus and to establish adequate intra-institutional backup to monitor, guide and implement projects in an effective manner.

The activities of the department were, therefore, extended to include a number of technology development programmes with the participation of the national laboratories of CSIR, the Indian Council of Agricultural Research (ICAR) and institutes of excellence such as the Indian Institutes of Technology (IITs). To meet the technological challenges, the department proposed an outlay of Rs 656 crore in the Eighth Five-Year Plan. The Planning Commission had, however, allocated only Rs 130 crore as an indicative outlay. By introducing a number of programmes oriented towards technology development, the department could get a total allocation of Rs 224 crore on my joining DOD during the last phases of the Eighth Five-Year Plan. This amount was more than double the outlay of the Seventh Five-Year Plan – Rs 108 crore.

On the international scene, the department took active steps to demonstrate its commitment to the principles enshrined in the UNCLOS and the Antarctic Treaty. India ratified the UNCLOS and the Implementing Agreement in June 1995, and the Environmental Protocol to the Antarctic

Treaty in April 1996. The country was accorded an important place in the Council of the International Seabed Authority (ISA) as a member of the investor category because of its deep commitment to the polymetallic (manganese) nodule programme.

The thrust areas during my tenure (1995–2001) included:

(i) contribution towards front-ranking research in polar sciences,
(ii) developments in technologies relating to seabed mining and extractive metallurgy,
(iii) exploration of marine living resources and studies to evolve strategies for their sustainable harnessing and utilization,
(iv) developmental activities related to an integrated coastal zone and marine area management,
(v) coastal community development through societal programmes,
(vi) capacity building through creation of centres of excellence in ocean science and technology in nine selected universities in maritime states, and
(vii) support for basic research in ocean science and technology.

Seven successful expeditions to Antarctica were launched between 1995 and 2001, of which six were to the Antarctic continent to pursue ongoing scientific activities. The seventh was a unique cruise through the Antarctic waters (Southern Ocean) for krill exploration and resource assessment. The country regularly took an active part in the policy formulation and implementation processes of the two international frameworks UNCLOS and Antarctic Treaty during this period.

The National Institute of Oceanography (NIO) at Dona Paula, Goa, established in January 1966, is the premier institution in the country that promotes research in ocean science. It comes under CSIR of the Ministry of Science and Technology. I recognized that practically all the activities being implemented by NIO were within the mandate of DOD. Hence, I felt that NIO should be with DOD and be responsible for the promotion of all activities related to ocean science in line with the national ocean policy. From the records available at DOD, I noted that there had been a decision in the early 1990s, at the level of the minister for science and technology, to transfer NIO to DOD but the then director general (DG) of CSIR did not implement it.

This information prompted me to take up the issue with the Committee of Secretaries (CoS), the meetings of which were held under the chairmanship of the cabinet secretary. However, I was not successful, despite two attempts. This setback was solely because of the unwillingness on the part of Dr R. Mashelker, DG of CSIR to part with NIO. However, in the final meeting of the CoS, Cabinet Secretary Prabhat Kumar opined that there was a need for better coordination between DOD and NIO. He decided that the secretary in the DOD should be the chairman of the NIO Research Council, which had the mandate to discuss and approve the R&D programmes in ocean science without any overlap with those implemented directly by DOD. Moreover, the DG, CSIR, should get the concurrence of the secretary to the DOD to depute NIO officials to international meetings on ocean programmes.

I understand that a proposal to transfer NIO to DOD was being actively considered by the Government of India even in 2018. However, NIO continues to be under CSIR even today.

Ninth Five-Year Plan Initiatives (1997–2002)

Well before the constitution of the working group for the formulation of the Ninth Five-Year Plan (1997–2002) programmes, I had thought of developing a ten-year perspective plan for oceans that covered a broad spectrum of activities. Hence, I constituted various expert groups to work on divergent subjects – with eminent scientists as chairmen and members, as outlined below. The expert groups were asked to identify future programmes of the department in a broad perspective by taking into account the importance of oceans as a resource for the growing population of the country and the need to protect and preserve the marine environment. The expert groups were:

Polar Science	Dr U.R. Rao
Marine Living Resources	Dr E.G. Silas
Marine Living Resources: Drugs and Chemicals	Dr S. Varadarajan
Marine Non-Living Resources: Minerals and Energy	Dr V.K. Gaur
Marine Environment and Coastal Zone	Dr A. Ramachandran
Ocean Atmosphere and Climatology	Dr R. Narasimha

Based on the inputs given by the expert groups, I identified two sets of activities for the Ninth Five-Year Plan:

- ongoing programmes with continued or increasing activity schedule, and
- initiatives in areas of ocean research and technology development for direct benefit to targeted sections of society.

The working group for the formulation of the Ninth Five-Year Plan proposals for ocean development projected a budgetary support of Rs 900.40 crore for the plan schemes. The Planning Commission, however, allocated an indicative budgetary support of Rs 557.70 crore to begin with, which was reduced and fixed at an outlay of Rs 510.62 crore. In spite of the fact that the Ninth Plan outlay was two-and-a-half times the Eighth Plan expenditure, a number of programmes prioritized for it either had to be deferred or implemented with restricted targets and objectives, as the approved outlay was hardly 57 per cent of the projected amount.

In line with the programme contents proposed by the expert group and approved by the Planning Commission, I took steps to reorient and restructure the department and the programmes to cover a wide spectrum of activities in the ocean sector, with emphasis on deliverables for the socio-economic benefit of the community and the nation. My first action was to reorganize the scientific wing at the DOD headquarters in New Delhi on the lines of ISRO, by adopting a programme-oriented approach. The programmes were as follows:

- Planning, Evaluation and Coordination (PEC),
- Bilateral and International Cooperation (BIC),
- Ocean Observation and Information (OOI),
- Polymetallic Nodules (PMN),
- Marine Research and Capacity Building (MRCB),
- Antarctic Expedition and Research (AER),
- Marine Living Resources (MLR),
- Marine Non-Living Resources (MNLR),
- Marine Environment and Coastal Zone (MECZ),
- Island Development, and
- Shore to Fishing Vessel Communication, among others.

In order to encourage focused programme implementation and the achievement of goals, I assigned responsibilities to my senior colleagues at the level of directors and principal scientific officers – Dr V. Sampath, B.N. Krishnamurthy, H.P. Rajan, Dr B.R. Subramanian, S.K. Das and Dr V.S. Chintala Rao. They were ably assisted by dedicated junior colleagues – Dr A. Mitra, Dr M.P. Wakdikar, Dr K. Somasundar, T.V.P. Bhaskar Rao, Dr P.G. Dastidar and Dr P. Madeswaran.

I also established three autonomous institutes under the management of the department – the National Institute of Ocean Technology (NIOT) in Chennai, the National Centre for Antarctic and Ocean Research (NCAOR) in Goa and the Indian National Centre for Ocean Information Services (INCOIS) in Hyderabad. In addition, I instituted attached offices of the department – the Integrated Coastal and Marine Area Management (ICMAM) Project Directorate in Chennai and the *Sagar Sampada* Vessel Management and Marine Living Resources Coordination Cell in Kochi, which was subsequently re-christened as the Centre for Marine Living Resources and Ecology (CMLRE). These innovative measures had a tremendous impact on streamlining the implementation of the department's programmes and in improving the outreach of the ocean development activities. I acknowledge the contribution of my colleagues and their teams in the fruitful realization of these efforts. A brief presentation of the performance of the department during my tenure is presented in the following six chapters.

17

The Indian Antarctic Programme

INDIA ENTERED THE ANTARCTIC realm by launching the country's maiden expedition to the ice-covered continent from the shores of Goa in December 1981. In 1983, India acceded to the Antarctic Treaty System (ATS) and established its first manned permanent research base at Dakshin Gangotri in Antarctica the following year (Photo No. 60). All activities of ATS are executed with the consensus of all member states who have the consultative status. Our country also attained consultative status in 1983. (It is one among twenty-nine nations to date.) Later, India became a member of the:

- Council of Managers of National Antarctic Programmes (COMNAP),
- Standing Committee on Antarctic Logistics and Operations (SCALOP),
- Commission for the Conservation of Antarctic Marine Living Resources (CCAMLR),

- Scientific Committee on Antarctic Research (SCAR), and
- Committee on Environmental Protection (CEP).

In 1988, Maitri, India's second manned permanent research station, was established at Schimacher Oasis in Central Dronning Maudland, Antarctica (Photo No. 61). It could accommodate sixty expedition members (twenty-five at the main station and thirty-five in summer camps). Dakshin Gangotri got buried in ice and was subsequently abandoned in 1990–91.

Annual Antarctic Expeditions

From 1981, DOD used to organize an Antarctic expedition every year. In order to catalyse and consolidate the gains accrued from the various expeditions, the Antarctic Study Centre (ASC) was established in 1988 in a small office in a rented building at Vasco da Gama, Goa. T.V.P. Bhaskar Rao was the head of ASC. The joint secretary (Administration) at DOD and the director of NIO, Goa, guided him as he made all the logistic arrangements for every expedition. He interacted with institutions to arrange for the training, medical tests and protocols to be complied with for the expeditions. I appreciate his substantial work inputs despite minimal support from the department.

The ASC continued the coordination of all Antarctic expeditions until the National Centre for Antarctic and Ocean Research (NCAOR) was established as an autonomous R&D institution of DOD on 25 May 1998. I organized the inauguration ceremony for NCAOR on 5 April 2000. The centre was dedicated to the nation by Dr Murli Manohar

Joshi, the minister for ocean development, science and technology, and human resource development.

Every Indian expedition to Antarctica had two groups of scientists – the first team of twenty to thirty scientists carried out research during the Antarctic summer, and the second set of eight to twelve scientists conducted studies for longer periods, during the summer and the following winter at the Antarctic Research Station. This number was in addition to the logistics personnel assigned to cater to the station maintenance, its life-support systems, communications and all associated tasks. The Indian Antarctic expedition – both the summer and winter teams – normally reached the continent by the end of each year and the summer team leaving behind the winter team in Antarctica returned around late March the following year along with the winter team of the previous expedition who had in turn been left behind in Antarctica fifteen months ago.

The DOD used to charter an ice-breaker class vessel to take the scientists from Goa to Antarctica and to bring them back. The total period of deputation of the summer team was about 100 to 105 days. I observed that on an average, thirty-five to forty days were spent on board, including travel both ways, for every expedition. During this time, very little scientific activity could be carried out.

In consultation with Dr V. Sampath and Dr A. Mitra of the DOD-HQ and Dr P.C. Pandey, the director of NCAOR, I examined the possibility of reducing travel time by launching the expedition to Antarctica from Cape Town, South Africa. The belongings of the scientists and the equipment could be shipped from India to Cape Town instead. A detailed analysis of the cost involved in the Antarctic expedition, a major part of which went towards chartering an ice-breaker, led me to conclude that shifting the launch of the expedition from Goa to Cape Town could bring down the expenditure substantially as time spent on cruise would go down by around twenty days. I interacted with the authorities of the South African government and the South African National Antarctic Expedition (SANAE) in this connection. I also visited Cape Town with my colleagues to finalize the infrastructure and logistics requirements. Based on this, and with support from the South African authorities, I prepared a detailed proposal justifying the plan and got the necessary approvals from the Indian government for the launch of expeditions to Antarctica from Cape Town.

All infrastructure facilities and logistics support were established with the guidance and assistance of the South African government. As planned, an official Indian Antarctic Expedition inaugural function was organized on the cruise, which was set to depart, at Cape Town on 9 December 1999. It was attended by Mohammed Valli Moosa, South Africa's minister for environmental affairs and tourism, and Harsh Bhasin, the Indian high commissioner in South Africa. Around midnight, Dr P.C. Pandey flagged off the expedition. India has since then been regularly launching the annual Indian expeditions to Antarctica from Cape Town as a routine operation.

I understand that about 3,000 km east of Maitri, a new Indian research base named Bharati was commissioned on 18 March 2012 to facilitate year-round scientific research activity in the Indian Antarctic programme (Photo No. 62). This facility can support forty-seven personnel on a twin-sharing basis in the main building along with an additional twenty-five people in emergency shelters/summer camps, thereby taking the total capacity to seventy-two.

Polar Science: Research Profile

The Annual Indian Antarctic expeditions have continued since 1981. Scientists with a keen interest in polar science, from over forty institutions, universities, research laboratories, government departments and so on, have been participating in the expeditions and pursuing research projects of their choice. However, I was dissatisfied because there was no coordination among the diverse research activities in multifarious areas and there was some avoidable duplication. Moreover, a system to monitor, review and guide Antarctic research was missing.

In 1995, I appointed a national-level committee under the chairmanship of Dr U.R. Rao, former secretary, Department of Space, with senior scientists and directors of laboratories engaged in polar science research as members. The objective of the National Expert Committee on Antarctic Research – as it was named – was to review research on polar science in the country, identify a programme profile for the upcoming decade and evolve a system, procedure and strategy to coordinate, monitor, review and guide its activities. The expert committee submitted its report in early 1996 and its recommendations were implemented following due procedure. Accordingly, I was able to restructure and reorient the polar research activities in Goa with the approval of the Government of India in 1997. I appreciate the contributions of the committee and thank everyone for their visionary ideas.

Many of the projects carried out in Antarctica are long term, while a few are for a shorter period of two to four years. The Indian Antarctic Programme has the following sections: atmospheric sciences, biological sciences, earth science and glaciology, environmental sciences, and human physiology and medicine, with climate change as the broader theme.

The National Centre for Antarctic and Ocean Research (NCAOR)

I observed that at the Expenditure Finance Committee (EFC) meeting of the department, held on 8 June 1988, approval was given to establish the Antarctic Study Centre (ASC) as an autonomous institute of DOD to stimulate research in the frontier areas of polar science and ocean sciences. ASC had been functioning as an office attached to DOD until then. Recognizing the urgent need for such a facility, I renamed ASC as the National Centre for Antarctic and Ocean Research (NCAOR) and registered it as an autonomous society under the Societies Registration Act of Goa on 25 May 1998 (Photo No. 63). NCAOR was established in order to consolidate the gains accrued from the Antarctic expeditions and to stimulate research in polar science and the Southern Ocean oceanography. In 2018, NCAOR was renamed again – called the National Centre for Polar and Ocean Research (NCPOR) – following its involvement in the scientific pursuits of both the Arctic and Antarctic regions.

As indicated earlier, the ASC was functioning in a rented building at Vasco da Gama. NIO in Goa had been guiding and helping it develop its infrastructure. About fourteen hectares of land at Headland Sada, Vasco, was allotted to DOD by the Government of Goa for the establishment of a full-fledged centre for Antarctic research. Unfortunately, this place was the garbage dump of the Vasco Municipality. DOD requested the government to issue the necessary orders so that the municipality would stop dumping garbage inside the new ASC campus and the existing garbage was shifted out of the land allotted. I even met the chief secretary of the government and explained the issue. His response was favourable, but the municipality remained non-cooperative. I asked the director of NCAOR to construct a compound wall for the premises to thwart the trespassing, and accepting my suggestion, he did build a wall. Regrettably, the municipal authorities broke the compound wall at night and again dumped the garbage. It thus became a legal issue and we approached the court, which issued orders in our favour. The garbage dumping was stopped; we secured the property and constructed buildings as planned.

During my tenure, DOD took over the responsibility of constructing the buildings for NCAOR from NIO. We erected a state-of-the-art, ultra-clean ice core laboratory for the archival work, and the processing and analyses of snow and ice samples from the polar region. The ice core laboratory was commissioned in January 2005. As a world-class and unique institution, it was equipped with updated ice core storage and processing facilities maintained at –20°C and –15°C, respectively.

The programmes pursued by the department over the years had also kept pace with developments worldwide and addressed national needs and issues based on the Ocean Policy Statement. I am pleased to record that NCAOR established a large number of facilities for polar and ocean science research in the disciplines of environmental studies, cryosphere and ice core studies (Photo No. 64), remote sensing, micropalaeontology, paleoclimatology, polar biology, oceanography, marine geophysics and hydrothermal studies among others.

60. Dakshin Gangotri: The first Indian Antarctic Station

61. Maitri Research Station: The second Indian Antarctic Station

62. Bharati: The third Indian Antarctic Station

63. National Centre for Antarctic and Ocean Research (NCAOR) main building, Goa, dedicated in 1998 to the nation by Dr Murali Manohar Joshi, minister for Ocean Development, GOI

64. Ice-core sampling in Antarctica

65. Sea trial of underwater sand mining system, a joint venture between the National Institute of Ocean Technology (NIOT) and University of Siegen, Germany, on board ORV *Sagar Kanya* in 2000

66. The Antarctic krill (Euphausiasuperba), harvested during Indian Antarctic Krill Expedition in 1995 from Indian Ocean sector of the Southern Ocean

Dr P.C. Pandey, a senior scientist from the Space Application Centre in ISRO, Ahmedabad, was appointed the director of NCAOR on 12 May 1997. He was largely instrumental in building the world-class infrastructure for polar research and logistics. Dr Pandey was supported by very efficient scientists – Dr A. Mitra, Dr S. Rajan, Dr M. Sudhakar and Dr N. Khare – and, at the DOD headquarters, by Dr V. Sampath and P.S. Singh, the section officer. Dr Pandey was given administrative, financial and procurement support by N.S. Dalvi, K. Sripati and K. Venkateswarlu, respectively. I congratulate all of them for demonstrating that an institute with an efficient and dedicated team of people can execute complex and challenging jobs and make the nation proud.

The First Antarctic Krill Expedition

The Antarctic krill (*Euphausia superba*) is central to the food chain of the Southern Ocean. The current annual catch (about 120,000 tonnes) is well below the precautionary catch limits set by CCAMLR. In view of the expanding market for krill and krill-based products and the perceived abundance of krill in the Southern Ocean, the interest in Antarctic krill fishery is expected to increase significantly in future. In 1995, DOD proposed to explore and commercially exploit the Antarctic krill in a sustainable manner. I recommended DOD's Fisheries Oceanographic Research Vessel (FORV) *Sagar Sampada* (71.5 m LOA; 2,285 HP; ice class IB) to be deployed for the purpose. Until then, the *Sagar Sampada* had not been taken out on any voyage to the Southern Ocean for survey or exploration. I decided to limit the duration of this first Indian Antarctic krill expedition to about ten weeks. I constituted an expert committee to evaluate the suitability of the *Sagar Sampada* for meeting the Antarctic krill expedition mission requirements and safety aspects. I also decided to equip the vessel for the proposed mission requirements.

Based on the recommendations of the expert committee, the FORV *Sagar Sampada* was prepared for the Antarctic krill expedition. A committee of experts on fisheries defined the objectives and drew up a detailed action plan. The expedition was launched on 27 December 1995 from the Kochi harbour, to obtain first-hand information pertaining to the harvesting and utilization of Antarctic krill resources, in Fishing Area 58 in the Indian Ocean Sector of the Southern Ocean.

The members of the first Antarctic krill expedition were selected from among those scientists working in diverse institutions on fisheries research. They also were associated with the DOD programmes. M.K.R. Nair, the director of the Integrated Fisheries Project (IFP), Kochi, was chosen as the cruise leader for the expedition. The five-member scientific team comprised the following personnel:

- M.K.R. Nair, cruise leader, with considerable expertise in fisheries harvest and post-harvest technology,
- Dr M.R. Boopendranath, Central Institute of Fisheries Technology (CIFT), deputy cruise leader and expert in harvest technology,
- A. Androse, expert in resource assessment working for the Fishery Survey of India (FSI),
- Dr C.N. Ravishankar, CIFT, expert in post-harvest technology, and
- Dr Vijayakumar Rathod, NIO, expert in biological oceanography.

Given that the expedition was the first Indian venture of its kind, I felt it wise to associate with the Sea Fisheries Institute (SFI), Poland, which had experience in Antarctic krill expeditions. Such an association enhanced our confidence in the success of the enterprise. I recalled the manner in which collaborations with expert agencies in France, Germany and Russia contributed to the success in many programmes of ISRO. I visited Poland, held discussions with the government and the SFI authorities and experts, and evolved an action plan for the expedition. The experts I met included the following, among others:

- Dr Piotr Jan Bykowski of SFI, an expert in post-harvest technology,
- Captain Roman Myczkowski of the Dalmor Deep Sea Fisheries Company, Gdynia, Poland, a specialist in polar navigation and harvest technology, and
- Dr Zygmunt Klusek of the Polish Academy of Sciences, Sopot, Poland.

Thus Poland actively participated in what came to be known as the First Indian Antarctic Krill Expedition (FIKEX). It had a fourteen-member Indian team that comprised fishing masters, technical assistants, processing assistants, service engineers, two representatives of the deep-sea fishing industry, fishing hands and thirty-five crew members from the Shipping Corporation of India (SCI), Mumbai.

The important objective of the mission was to gain experience in detecting and harvesting the Antarctic krill from the Indian Ocean sector of the Southern Ocean. During the expedition, well-defined krill swarms were located during day and night by echo sounders. The average catch rate of the Antarctic krill during the operation worked out to be 379 kg per haul. In the course of ten weeks, the vessel covered about 12,550 nautical miles. Approximately 12,070 kg of catch was landed by aimed mid-water trawling operations. The Antarctic krill constituted 47.1 per cent of the total catch, while salps made up 52.5 per cent of the landings.

After the successful completion of the mission, the FORV *Sagar Sampada* returned to the Kochi harbour on 10 March 1996. The Antarctic krill catch of over 5 tonnes that was utilized for (i) studies on biochemical and processing aspects at the Central Institute of Fisheries Technology (CIFT), Kochi, and (ii) for krill-based product development such as krill sausage, krill ball, burger, cutlet, kebab, chutney powder, pickle and smoked krill seasoning, krill meal, chitin and chitosan at IFP.

From the first Antarctic krill expedition, it emerged that high investment requirements, operational cost and issues in the processing and utilization of high fluoride content in the krill shell were problems to be resolved before undertaking commercial operations. I congratulate M.K.R. Nair and all the members of the expedition for their excellent contributions. I heard that the FORV *Sagar Sampada* once got caught in a heavy storm in the Southern Ocean but the passengers and the vessel had come out of the incident unscathed.

The Antarctic Treaty System

As secretary, DOD, who represented India, I participated in all the Antarctic Treaty Consultative Meetings (ATCMs), on a consultative status. During one of the ATCMs in Norway, I held discussions with the officials of the Norwegian Polar

Institute about the possibility of India initiating studies in the fields of Arctic microbiology, atmospheric sciences and geology at the International Arctic Research Facilities in Ny-Ålesund, Norway. The response was positive. In the course of the ATCM, I had the unique experience of seeing the sun day and night.

In 2007, I was delighted to learn that the dialogue between India and Norway progressed successfully and that India had begun its scientific endeavours in the Arctic. India had embarked on a long-term programme of regular scientific activities in the Arctic in the frontier realms of polar biology, glaciology and earth and atmospheric sciences. NCPOR is coordinating the Arctic research programme as well, with emphasis on climate change and some of the frontier areas of polar science of special relevance to the Arctic realm. To facilitate its activities, India had taken a station building named 'Himadri' on lease, at Ny-Ålesund, to serve as the research base in the Arctic. Himadri was inaugurated by Kapil Sibal, the minister for earth sciences, on 1 July 2008 in the presence of dignitaries from Norway, Germany and other nations involved in the Ny-Ålesund activities.

The year 1998 turned out to be a memorable one for India as far as the Antarctic research was concerned. For the first time after it became a member of the Antarctic Treaty System, India successfully hosted the Working Group on Ecosystem Monitoring and Management (WG-EMM) of CCAMLR at Kochi during 10–21 August 1998. The meeting was attended by the CCAMLR member countries. The first joint technical meeting on cooperation between India and Peru in Antarctic research was held at Lima during 3–4 August 1998. The two sides identified joint Antarctic research on geology, krill biology, upper atmospheric science and so on.

In May–June 1999, the second meeting of the Committee on Environmental Protection (CEP-2) of the Antarctic Treaty System was held in Lima in conjunction with the twenty-third ATCM. Dr V. Sampath of DOD and I participated in the meeting and expressed our concern about the increasing environmental degradation in the Southern Ocean and the Antarctic waters, which was due to the unbridled growth of Antarctic tourism. We requested the committee to recommend putting a cap on the annual tourist footfall on this pristine continent to minimize adverse environmental impacts. In 1999, New Zealand and the US sponsored a special meeting of the ministerial delegation that comprised all member states with consultative status to Antarctica, from 25 to 28 January 1999, in commemoration of the fortieth anniversary of the Antarctic Treaty System. Each member state nominated two representatives – one at the level of minister and the other a senior official. Prime Minister Atal Bihari Vajpayee – who was also the minister in charge of DOD could not attend. Therefore, I represented India in the special meeting in Antarctica.

The entire team assembled at Christchurch, New Zealand, and flew to the South Pole in two aircraft. Accommodation was provided in research stations around the South Pole. Apart from a meeting of all members, the participants were divided into groups and taken in helicopters to various tourist sites – the South Pole, Scott's Hut, Cape Evans, the research station in the Dry Valleys, Shackleton's Hut at Cape Royals, Penguin Colony at Castle Rock and McMurdo/Scott Base area, and the Italian station at Terra Nova Bay. It was a thrilling experience to live in the southern part of Antarctica for about a week. However, I could not visit Maitri, the

Indian Antarctic research station, since it was located far from where the delegation was stationed.

India Chairs the CCAMLR in Hobart, Tasmania

The Commission for the Conservation of Antarctic Marine Living Resources (CCAMLR) came into existence in 1982. Since then, the member states have been meeting in Hobart, Tasmania, every year. At present, the CCAMLR has twenty-four member states including India, plus eleven acceding states with a secretariat functioning from Hobart. The chair of CCAMLR is taken by the member states in rotation in alphabetical order for a period of two years. India became the chair in 1999 and 2000. Consequently, I discharged my responsibility as the chairman of CCAMLR, attending its meetings during these two years. It was an extremely valuable experience to chair an international meeting where every conservation measure and scheme had to be negotiated and adopted only after the acceptance and approval of all member states. Even if one member objected to a proposal, it had to be carried over to the next meeting for consideration. I observed that the member states were functioning in groups with their own agendas and strategies.

During the 1999 meeting, I faced two major issues – the first one concerned resolving a procedural issue related to the service of Esteban de Salas, executive secretary in charge of the CCAMLR office, and the other was an agreement on the text that made up the Catch Documentation Scheme (CDS), which had failed to obtain the consensus of all member states in the previous two years or more. I discussed these issues in the daily meetings of heads of states and also in my private conversations with the delegations of each member state. I shared the views of the various members with others and suggested a few solutions for a compromise. It was a moment of immense gratification when I finally succeeded in resolving these two issues to the satisfaction of all the members. The second meeting in 2000 was one of the most productive. The CDS underwent a comprehensive review and new texts (both conservation measures and resolutions) supporting it were adopted in that meeting.

The heads of many states appreciated the constructive, positive, and thoughtful way in which I conducted the meetings. Some members thanked me for my skilful chairing and the approach that sought the views of all parties to resolve issues satisfactorily. Dr V. Ravindra Nathan of the DOD Marine Living Resources programme, who associated with the CCAMLR meetings, was with me during the two years. I thank him for his valuable advice in arranging meetings with the heads of member states, which often helped arrive at a consensus.

18

The Marine Resources Programme

EXPLORATION AND EXPLOITATION OF marine resources in a sustainable manner is one of the thrust areas of the department. These marine resources are classified as non-living resources and living resources. The efforts involved are multidisciplinary and executed through multi-institutional participation and guided by experts in the country. In this chapter the marine non-living resources are presented followed by marine living resources.

Marine Non-Living Resources (MNLR) Programme

Polymetallic Nodules Programme

The Polymetallic Nodules (PMN) programme was one of the major R&D efforts of DOD – now the Ministry of Earth Sciences (MoES), Government of India – to explore and exploit the mineral resources in the deep sea beyond the national jurisdiction. This multidisciplinary programme was made possible by multi-institutional participation. The first sample of PMN was collected in 1981 from the Indian Ocean. Continued efforts succeeded in identifying a prospective site with PMN in the Central Indian Ocean Basin (CIOB). The discovery led to the subsequent recognition of India as a pioneer investor by International Seabed Authority (ISA) in 1987, along with Japan, France, Korea, the Inter-ocean Metal Consortium (Consortium of East European countries), China and the USSR.

India has the distinction of being the only country allotted the exploration area in the Indian Ocean when all others were apportioned similar expanses in the Pacific Ocean. Under this programme, the Indian efforts are directed towards survey and exploration, development of deep-sea mining technology and extractive metallurgy technology, and environmental impact assessment. It is planned to develop and demonstrate adequate technological capabilities in all the areas mentioned above, in the years to come, so that India can embark upon a commercial venture when it is economically viable. Initially, this programme was piloted by S.K. Das and subsequently by Dr M.P. Wakdikar of DOD. I appreciate their sustained efforts in this venture.

Survey and Exploration

The first and major component of the PMN programme was survey and exploration, aimed at firmly establishing the resource potential. NIO of the CSIR and the NCAOR of DOD contributed to these efforts. India has made considerable progress in resource identification, assessment and associated technology and manpower development. Our country was accorded the status of Pioneer Investor and 150,000 sq. km was allocated in the CIOB by the Preparatory Commission of the UN before the establishment of ISA for exploration and exploitation studies of the resources.

As per the provisions of the United Nations Convention on the Law of the Sea (UNCLOS), before the commencement of the mining of nodule resources, the Pioneer Investor – India – had to fulfil various obligations. One of the requirements was to relinquish 50 per cent of the area in three phases. India had already completed the topographic survey of the pioneer area. The abundance pattern vis-à-vis the geomorphology had been established along with an upgradation of the resource evaluation techniques. Moorings deployed in the pioneer area obtained continuous time series oceanographic data. The vast data thus acquired provided new insights into the topography of the basin and the resource potential. Further, closer grid surveys at 5 km x 5 km were carried out and deep-tow photography was done in selected areas to improve the estimates on resource potential and demarcate the mine site and mining tracks. Based on these surveys, India relinquished 75,000 sq. km of area in three stages until March 2002 and signed a new fifteen-year contract with ISA for the 75,000 sq. km retained. Dr M. Sudhakar piloted the survey and exploration activities in an exemplary manner.

Deep-Sea Mining Technology

The second component of the PMN programme was development of deep-sea mining technology. The design and development of a remotely operated, underwater mining system with collector module, lifting module, and instrumentation and control systems were initiated in 1990 at the Central Mechanical Engineering Research Institute (CMERI), Durgapur. The preliminary attempts were to test the concepts of deep-sea mining, to generate some of the basic data and to understand the functional and operational needs of the system and its subsystems. The initial design of the underwater mining system with a capacity of 100 tonnes/day had a remotely operated, crawler-based collector module, a bucket-in-pipe–based lifting module and control module with the instrumentation required. The performance of the subsystem was evaluated on land and subsequently in an on-shore shallow test basin. Concurrently, the development of a Remotely Operated Vehicle (ROV) for the inspection and maintenance of the underwater system was also undertaken. The aforesaid activities were formulated to generate data for a mining system that was capable of operating up to a depth of 5,000–6,000 metres.

In 1995, I got the accomplishments of CMERI in the field trials of these subsystems on land and in a shallow basin evaluated by an expert committee. I also studied the international scenario in this field and felt that there was much to be gained from appropriate technical collaboration with institutions abroad with the requisite experience and expertise. I assigned the responsibility of developing deep ocean mining technology to DOD's National Institute of Ocean Technology (NIOT). These developments had two components – the first, an ROV for underwater observations and the second, equipment/machinery for seabed mining.

During 1996–97, the ROV 200 was sponsored by DOD at the CMERI to provide underwater observations up to 200 m depth. NIOT was associated with this development. The system was developed, transported and mounted over a floating barge at Jamnagar Port for demonstration and field trials. However, it was severely damaged in the cyclone in Gujarat in June 1998. Corrective actions took over three years and a successful sea trial was carried out in August 2001 by the CMERI and NIOT at a depth of about 190–195 metres.

Based on the outcome of the demonstrated technology, two projects – the ROV 500 with a higher depth capability up to 500 m and the Autonomous Under water Vehicle, AUV 150, a collaboration between CMERI and IIT Kharagpur – were initiated in May 2003. I understand that the sea trial of the AUV-150 was conducted by CMERI from the Chennai coast up to 150 m depth with active support from NIOT in August 2011. The successful sea trial of ROV 500 up to 505 m depth was completed in August 2013 along the same coastline. Further, in order to supplement the technology development efforts for underwater observations, I made unstinted efforts for joint projects with the Russian Academy of Sciences (RAS).

Towards the end of my tenure in DOD, with the approval of the Government of India, I reached an understanding with the Experimental Design Bureau of Oceanological Engineering (EDBOE) of the Russian Academy of Sciences for the joint development of an ROV with NIOT to operate at 6,000 m depth. This step resulted in an agreement/MOU between EDBOE/RAS and NIOT/DOD in 2003. EDBOE was identified as a development partner in 2005. I am happy to note that NIOT carried out the developments and successfully demonstrated the deep-water ROV in field trials at CIOB at 5,389 m depth in 2010.

Before undertaking developments for mining PMN from 6,000 m depth, I planned a demonstration of shallow-water sand mining technology in collaboration with the University of Siegen, Germany, which had experience in underwater mining systems. In 1998, an MOU was signed between Das Interne Kontrollsystem (IKS), University of Siegen, and the NIOT/DOD for the design, development and demonstration of underwater sand mining system at 500 m depth in Indian seas and the preparation of a detailed project proposal for mining of PMN from 6,000 m depth in the CIOB.

Under this joint project, the entire sand mining system – with sand cutter, manipulator, slurry pump and flexible riser system – was built jointly by the University of Siegen and the NIOT, except the undercarriage, which was made by IKS. Four sea trials were carried out on-board the ocean research vehicle ORV *Sagar Kanya*, with a successful demonstration on 14 March 2000, off the Thoothukudi coast at 410 m depth (Photo No. 65). Further technology developments were carried out by the NIOT to meet the DOD programme needs of the future.

Extractive Metallurgy Technology

The third component of the PMN programme was the development of extractive metallurgy technology to produce minerals and metals from these nodules. DOD established a committee of experts under the chairmanship of Dr C.V. Sundaram, the group director of BARC in the Department of Atomic Energy, to guide, monitor and periodically review the progress. I congratulate Dr C.V. Sundaram and all the members of his committee on their contributions.

Copper, nickel, cobalt and manganese are the major metals of interest obtained from polymetallic nodules. In 1984, certain national laboratories and industries started R&D for the recovery of three metals – copper, nickel and cobalt – from the nodules mined from the Indian Ocean. After studying and analysing the merits of about fifteen process routes, India chose the hydrometallurgical option. In late 1980s, the Regional Research Laboratory (RRL) in Bhubaneswar, the National Metallurgical Laboratory (NML) in Jamshedpur and the Hindustan Zinc Limited (HZL) in Udaipur intensified their process development efforts through the ammoniacal sulphur dioxide leach route, roast reduction ammoniacal leach route and acid leach pressure-leach route, respectively. In addition, the metallurgy group of BARC was engaged in developing a process to extract cobalt from the nodules. Over the years, more than 50 tonnes of nodules from the Indian mine site have been processed by these institutions.

In the latter half of 1995, taking into consideration the economic advantage and the status of recovery of metals from the nodules, I reoriented the extractive metallurgy programme to recover the compounds of manganese, in addition to copper, nickel and cobalt. Thus, the PMN programme was directed towards the recovery of four metals.

I also observed that the research efforts in the laboratories were open-ended. It was necessary to focus on R&D and establish a production unit in an industry with the necessary infrastructure. Therefore, I requested RRL, NML and HZL to submit a detailed report on their achievements and plans.

This report was reviewed by the PMN Management Board. Based on the recommendations of the board, I requested all the three agencies to carry out additional experiments within the next three months and to submit a comprehensive report on the development. This report was also presented to the PMN Management Board. I requested the board to review the status and recommend one process to establish a pilot plant in an industry. Meanwhile, I visited HZL in 1996, saw their facilities in Udaipur and discussed with its chairman and managing director (CMD) the possibility of establishing a continuously operating pilot plant with a capacity to process 500 kg dry nodules per day, based on the technology selected by DOD, with continued support from the developing agencies. He expressed his willingness to proceed with the project.

In 1997, the result of the batch scale process development was reviewed by the PMN expert committee and it was decided to develop the sulphur dioxide ammoniacal leaching process further in a continuously operating pilot plant with a capacity to process 500 kg dry nodules per day. R&D efforts progressed at the Indian research institutes to generate data to facilitate the design and establishment of a continuously operating pilot plant at HZL. Based on the agreement executed between DOD and HZL, I laid the foundation for the pilot plant on the HZL premises for a semi-continuous pilot demonstration plant.

I am glad to note that this plant was commissioned in 2002 to produce copper, nickel, cobalt and manganese from PMN mined from the CIOB. I am given to understand that successful campaigns were carried out till 2006, based on metallurgical process routes of RRL (now redesignated as the Institute of Minerals and Materials Technology), NML and HZL, with the support and guidance of BARC. Flowsheets for extractive metallurgy were established. The venture laid the foundation for the subsequent metallurgical extraction programme of MoES for copper, nickel, cobalt and manganese from polymetallic nodules.

Environmental Impact Assessment of Deep-Sea Mining

The fourth component of the PMN programme was environmental impact assessment (EIA) studies. It was recognized that the mining of deep-sea minerals adversely affected marine ecosystems. Hence, studies in connection with the mining of deep-sea PMN were undertaken to evaluate the possible environmental impacts of the process to fulfil one of the obligations of the country as a pioneer investor under UNCLOS. Baseline environmental data on various biological, physical, chemical and geological parameters were collected, along with data from the test and reference sites. A benthic disturbance was artificially created in 1996–97 in the allocated site of the CIOB to simulate the impact of deep-sea mining.

In order to study the effects of sediment resuspension and resettlement, the environmental parameters were monitored every year by collecting samples from the test and reference areas. In 2004, the results of the monitoring cruises indicated that the benthic conditions were steadily moving towards restoration and the effect of the disturbance was waning off. Further, an assessment of nodule-associated fauna and other environmental parameters was carried out after 2002. Studies conducted so far have resulted in understanding the baseline conditions and the environmental variability of different ecological parameters associated with deep-sea minerals. No adverse impact has been reported.

Polymetalic Sulphide Exploration

The concerted and sustained work undertaken by the Government of India in the course of three decades and my efforts to promote the extension of the PMN programme provided the lead and focused direction to extract four metals from the PMN mined from the Indian site in the CIOB. Further, based on the exploratory cruises in the south-west Indian Ocean, India got approval from ISA for a new area of 10,000 sq. km for polymetallic sulphide exploration near Rodrigues Triple Junction of the Central Indian Ridge and the South-west Indian Ridge. Thus, the Indian programme on deep-sea minerals now includes polymetallic sulphides and polymetallic manganese nodules.

The Marine Living Resources (MLR) Programme

The exploration of marine living resources and studies to evolve strategies for harnessing them sustainably are two of the objectives of DOD. The aim is to develop and use science and technology for the benefit of the coastal community and the nation. India has a coastline of about 8,000 km and a large human population lives within 50 km of the coast. Marine living resources are a vital source of protein for many of these communities. However, the trend in the marine fisheries annual catch in the recent past raises issues of sustainable exploitation and the scientific knowledge necessary to ensure yield. Intensive studies ought to be done on the size of the fish stock, the ecological role of individual fish species and their migratory and breeding patterns.

The predictive models on fishery resources should take into account the interaction between various fish species and integrate oceanic parameters, such as temperature, salinity and hydrodynamics, nutrients and food organisms. Small-scale fisheries may be less important in terms of macroeconomics, but are extremely vital for food supply and the employment of coastal populations. It is seen that the fish catch beyond 50 m depth is considerably small compared to the coastal fishery, and hence, the need for developments in deep-sea fisheries.

In 1995, I constituted an expert group on marine living resources to formulate a ten-year perspective plan to identify relevant programmes on the topic under the chairmanship of Dr E.G. Silas, former director of CMFRI and vice chancellor of Kerala Agricultural University, as a prelude to formulating the Ninth Five-Year Plan proposals. I inducted Dr V. Sampath of DOD in the expert group to provide inputs from ICAR, GOI, on the basis of the experience he had gained in marine fisheries development while working in the fisheries division of the Ministry of Agriculture.

The expert group recommended – keeping in view the potential for exploration and exploitation of marine living resources and the present status of R&D activities in the field of stock assessment, utilization of the resources, mariculture and other associated activities – a number of major thrust areas for future development. It also took note of programmes implemented/proposed by other ministries/ departments, such as the Ministry of Agriculture and the Ministry of Food Processing Industries, so as to avoid overlap or duplication of activities. Moreover, the decision taken at the meeting of the Indian Ocean Fishery Commission held during 15–19 January 1995 at Jakarta – in which Dr V. Sampath represented India from the Ministry of Agriculture – emphasized a realistic assessment of the marine living resources beyond the 70 m depth zone, which hitherto had not been given adequate importance. The major thrust areas recommended by the expert group were:

(i) Assessment of marine living resources in the Indian Exclusive Economic Zone (EEZ) in correlation with oceanographic parameters to increase marine productivity;

(ii) Development of modern technologies in mariculture for industrial and societal application by setting up pilot projects;

(iii) Marine ornamental fish culture;

(iv) Seaweeds and their commercial use; and

(v) International cooperation, including liaison and joint ventures.

Some of the strategies outlined by the expert group for sustainable development, management, conservation and exploitation of MLR were:

- Uniform methodologies for the assessment of living resources;
- Collection, compilation, analysis, interpretation and dissemination of data on the resources in the Indian EEZ for the use of the fishing community and the fishing industry;
- Enhancement of MLR through sea ranching, upgradation of harvest and post-harvest technology, and product development;
- Mariculture of commercially important finfish and shellfish species and demonstration of the viable technologies and ecologically sound package of practices for societal and industrial application;

- Demonstration of the technology for breeding, culture, transportation and marketing of commercially important marine ornamental fishes through pilot projects;
- Development of technologies for commercial cultivation of seaweeds and their processing; and
- Joint-venture fishing activities in the southern ocean and product development out of the Antarctic krill, cephalopods and fish.

The group also strongly recommended the creation of an exclusive central project implementation unit at the DOD headquarters to coordinate, monitor and review the implementation of these programmes and position DOD personnel at the respective national research institutes at the field level to achieve the targets and objectives. For the smooth implementation of the aforesaid programmes, the group recommended an outlay of Rs 400 crore for a ten-year period, of which Rs 223 crore was projected for the Ninth Five-Year Plan period (1997–2002).

The FORV *Sagar Sampada* was the only multidisciplinary vessel of the department, which undertook sixty-eight cruises – including the one to the Antarctic waters – between April 1991 and May 1996 for studies on the aforesaid disciplines. Approximately 1,144 participants, including foreign participants, took part in these cruises. The facility for on-board data collection was extended to twenty-four organizations, including universities, and 190 participants per annum utilized this facility. The cruises have, therefore, helped in identifying the location for underexploited and potentially rich deep oceanic resources. Some of the major achievements on the resources survey, oceanographic studies and ecosystem dynamics were:

- An acoustic survey to obtain data on the relative abundance of biomass in selected areas;
- Identification of the spawning and nursery grounds of mackerel, oil sardine, threadfin bream, ribbon fish, cat fish and so on;
- Identification of the areas with potential resources of deep-sea species such as the Indian driftfish, bullseye, pink perch, ribbon fishes, carangids, deep-sea prawn, deep-sea lobster and cuttlefish;
- Initiation of studies on ecosystem dynamics; and
- Identification of hitherto unexploited deep-sea fishery resources.

The seventy-five-day Antarctic cruise on the FORV *Sagar Sampada* that I had organized during December 1995–March 1996 to study the potential of krill (Photo No. 66) in the Indian Ocean area of the Antarctic waters covered Fishing Area 58 between 57°53' to 61°15' southern latitude and 31°40' to 36°31' E longitude. The following significant scientific objectives were achieved during this cruise:

(i) Physico-chemical and biological processes in relation to krill swarming;

(ii) Phyto-plankton and zooplankton and their relation to krill distribution;

(iii) Identification of krill aggregation and its characteristics by acoustic and visual means, and an assessment of krill harvesting potential; and

(iv) Appraisal of the distribution and abundance of krill, cephalopods, Antarctic fish, etc.

Studies on the commercial feasibility of exploitation of krill and other Antarctic living resources continued. Aspects of post-harvest technology of krill meat processing and the associated biochemical and post-mortem changes, storage and product development – including the preparation of ready-to-cook and ready-to-eat products from krill and Antarctic fish meat – were diligently examined. The extraction of fish protein concentrate and the derivation of chitin and chitosan from krill shell in the shore laboratory were also undertaken.

Centre for Marine Living Resources and Ecology (CMLRE)

To develop effective management policies and practices for the judicious exploitation of living resources, national research institutions need access to information on multi-species models for divergent resource systems, which can be applied to local conditions. Developing models of this kind is the focus of high-priority research at the international level. An evaluation of the information currently available at various national institutes involved in marine living resources research is essential to achieve these objectives. This knowledge could be correlated and corroborated with fishery dependent and independent factors to evolve diverse multi-species models for the sustainable development and management of living resources. Such a study would also help us implement regulatory measures for the conservation and management of these resources.

I felt that the excellent work being carried out by many institutions lacked focus and that an integrated effort with a mechanism for the generation of data and to coordinate and facilitate the monitoring, control, surveillance and management of our resources was necessary. During my tenure, the Sagar Sampada Cell at Kochi was declared as an attached office of DOD with effect from 26 December 1995 with V. Ravindranathan in charge. It was renamed the Centre for Marine Living Resources and Ecology (CMLRE) on 12 January 2001 and Ravindranathan became its director. The centre aimed to work closely with the marine research group of the Cochin University of Science and Technology (CUSAT), Kochi, the numerous fisheries research institutions located in Kerala and quite a few experts settled in the city after superannuation, who had expressed an interest to continue marine research.

I held discussions with the vice chancellor of CUSAT and senior professors – including Prof N.R. Menon, Prof K. Damodharan and Dr E.G. Silas, the former director of the Central Marine Fisheries Research Institute (CMFRI) and scientist-in-charge of NIO-RC (regional centre) in Kochi – regarding establishing the Centre for Marine Living Resources and Ecology (CMLRE) on the CUSAT campus. In 1998, I proposed that DOD construct a multi-storey building to establish office and research facilities there and share it with the marine research teams of CUSAT and other volunteer fishery experts. Everyone concerned agreed to it in principle. However, the proposal was dropped since it could not be carried through in the CUSAT senate.

Subsequently, I approached the government of Kerala for suitable land for the CMLRE. I am glad to know that the Kerala government subsequently allotted about 10 acres of state-owned land at LNG Road, Puthuvype, in Kochi, on 26 March 2008. Over the years, the team has established infrastructure facilities and research laboratories, which include the instrumentation lab, dry chemical lab, wet lab, plankton analytical lab, fishery lab, census of marine life lab, plankton wet lab, microscope lab and processing and computational lab among others. After the retirement of V. Ravindranathan, Dr V.S. Sanjeevan and Dr P. Madeswaran becme directors in that order. Dr M. Sudhakar took charge as director in August 2015. On 23 February 2019, I participated in the formal opening ceremony of the CMLRE on the new campus. It was inaugurated by Dr Harsh Vardhan, minister of earth sciences, who named it Atal Bhavan, CMLRE.

Drug from the Sea

India is endowed with a rich marine biota along its 8,000 km coastline, comprising over 500,000 species. It is recognized that marine flora and fauna possess diverse bioactive substances. Therefore, the oceans are also potential sources of a wealth of pharmaceutical products. Academic and pharmaceutical groups have equally focused their efforts on the isolation and characterization of biologically active compounds derived from marine organisms. These compounds encompass a wide spectrum of chemical structures, the majority of which are small molecules ideally suited to serve either as themselves or through chemical modification as new pharmaceutical agents for the treatment of life-threatening diseases such as cancer.

The National Project on Drugs from the Sea was taken up by DOD in 1990–91, under the auspices of CSIR's Central Drug Research Institute (CDRI), Lucknow. Dr S. Varadarajan, former DG of CSIR, had been guiding these studies and assisting DOD with implementing these projects. About fourteen national/state R&D laboratories, including those from universities, actively participated in the bio-prospecting of marine organisms in the Indian waters for drug discovery and drug development, using state-of-the-art facilities.

More than 4,000 samples collected from about 1,400 species of marine flora and fauna have been evaluated for their biological activity. Several of these samples were found to exhibit mild to strong therapeutic, anti-diabetic, antidiarrhoeal, anti-hyperlipidemic, anti-anxiety, anti-cholesterol-cum-antioxidant, antibacterial-cum-anti-fungal and larvicidal activities. Out of the six marine organisms that showed these activities, substantial progress has been made in the development of three samples – anti-diabetic, anti-diarrhoeal and anti-hyperlipidemic – which were taken up for toxicity studies. Their activities were revalidated in repeat collections, and regulatory subacute toxicity studies were done on two rodent species. The rodent toxicity study was completed and the samples were found safe and devoid of any side effects. The regulatory subacute toxicity studies of the anti-diabetic and antidiarrhoeal preparations in primates were also completed.

I realized that bringing a drug into the market is a very complex task; several mandatory stages have to be successfully completed and the process is time-consuming. I felt the need to undertake the exploration, identification, extraction, bio-evaluation, recombinatorial chemical analysis and synthesis of active compounds up to product development with the participation of pharmaceutical industries concerned and to work in 'mission mode'. The need to integrate and focus the efforts undertaken by divergent research institutions and pharmaceutical industries was also paramount.

In 1996, along with Dr S. Varadarajan, I visited the CDRI facilities in Lucknow and discussed the matter with its director and senior scientists there. The CDRI system and procedure for developing drugs from marine organisms was upgraded. In spite of the best efforts of all the participant institutions, CDRI could not bring to the market even one drug during my tenure at DOD. The complexity of the procedure and the inherent variations in the activity levels in species taken from the ocean – due to several factors, including the environment – perhaps account for the situation.

To my understanding, CMLRE has continued to participate actively in the drug development programme, with cruise support from the FORV *Sagar Sampada*. A Centre of Excellence (CoE) on drugs from sea was established at CMLRE during the Twelfth Five-Year Plan period and the infrastructure facilities were strengthened at NIOT in Chennai, including submersibles and deep-sea samplers. A team of experienced and dedicated scientists, including taxonomists, were engaged full-time in various activities – from the collection of marine samples to product development with industrial partners. Over the years, the Drugs from Sea project has progressed quite well, integrating the efforts of both R&D institutions and important pharmaceutical industries. I understand that four drugs from marine organisms are at present undergoing clinical trials. They are:

(i) Anti-hyperglycemic product (CDR-134-D-123),

(ii) Anti-hyperlipidemic product (CU1-002),

(iii) Antidiarrhoeal product (CDR-134-D125), and

(iv) Anti-hyperglycemic-cum-anti-hyperlipidemic product (CDR-134-F-194).

I appreciate the leadership of the CDRI and the exemplary work of scientists from a large number of research laboratories in our country. I would also like to thank Dr S. Varadarajan for his guidance and support in this national project and those involved at DOD during my tenure – especially Dr V. Sampath and Dr P. Madeswaran – for coordinating the efforts of the CDRI, other research laboratories and academia involved.

19
Ocean Observation and Information Services (OOIS)

THE GENERATION AND DISSEMINATION of reliable coastal and ocean data on waves, winds, temperature, current, salinity, upwelling, potential fishing zone information, coastal maps and so on are essential to support coastal and offshore developmental activities and further quality oceanographic research. Recognizing this need, DOD initiated the Marine Satellite Information Service (MARSIS) programme in 1990, to develop capabilities for the utilization of the remote sensing data of ocean and coastal zones, which would, in turn, help in the generation of useful ocean data and their dissemination to end users on an operational basis. Five MARSIS centres were set up on the east and west coasts of the country to interact with the downstream users. In addition, the deployment of drifting buoys for the collection of in-situ met-ocean (meteorology and [physical] oceanography) data and the development of coastal zone information systems were envisaged.

In 1992, DOD initiated a sea level monitoring and modelling (SELMAM) programme to assess the variations in the sea level due to climatic changes and other factors, and their impact on the coastal belt of India. For this purpose, tide gauges were installed near major ports and elsewhere in the country for the accurate measurement of tidal data. Since 1993, DOD has implemented the Joint Global Ocean Flux Study (JGOFS), a core component of the International Geosphere–Biosphere Programme (IGBP), designed to determine carbon flux in the world oceans. The purpose of the JGOFS programme (1987–2015) was to establish the role of oceans in governing the atmospheric carbon dioxide content, in relation to climate change. The field data generated in this programme were utilized to develop a series of coastal and ocean models.

In 1995, I observed that a large volume of data was generated by different agencies engaged in oceanographic research in India. It was necessary to coordinate these efforts and integrate them to generate 'data products', which would find applications in several operational systems. In 1997, I restructured and integrated all these programmes into the Ocean Observation and Information Services (OOIS). OOIS comprised four components: (i) Ocean Observing Systems (OOS), (ii) Ocean Information Services (OIS), (iii) Satellite Coastal Oceanographic Research, and (iv) Ocean Modelling and Dynamics. The primary objective of the OOIS programme was to generate reliable coastal and ocean data products pertaining to winds, waves, sea surface temperature, currents, upwelling, potential fishing zone forecasts, coastal maps and so on, for potential use in coastal and offshore development activities and to further the quality of oceanographic research.

Ocean Observing Systems (OOS)

The objective of the Ocean Observing Systems (OOS) programme is to acquire in-situ and remote sensing oceanographic and surface meteorological parameters on a real-time basis from the seas around India. The parameters being measured under the programme include surface winds, waves, atmospheric pressure, atmospheric temperature, sea surface temperature, salinity, sea temperature profiles, dissolved oxygen, hydrocarbons, nutrients, radioactivity, sea levels and so on. The instruments deployed to measure these parameters are moored data buoys, drifting buoys, expendable bathythermographs (XBTs), current meter arrays and tide gauges. Ocean observations through remote sensing by satellites would be validated using the sea truth data generated under this programme. The OOS project consists of three sub-projects: National Data Buoy Programme (NDBP), Sea Level Observation, and Drifting Buoy and other Observing Systems.

The National Data Buoy Programme (NDBP)

Recognizing the need for uninterrupted, real-time oceanographic data for the safety and economic development of the coastal community, the Government of India assigned the responsibility for Instrumented Data Buoy–based field observations to DOD and satellite-based remote-sensed observations to DOS in 1993. In 1996, I established the National Data Buoy Programme (NDBP) (Photo No. 73) at

NIOT. NDBP designed a system by selecting the best among all available subsystems and components. Oceanor and Seatex are two reputed suppliers of moored instrumented ocean buoys. Both responded to our public tender twice. I also learnt about the negotiations between them for merger. I selected the better of the two products on offer, negotiated with the representatives of both, and settled for the lower price of the two offers for every product, and also arranging the transfer of technology.

Subsequently, Seatex merged with Oceanor. DOD entered into an agreement with Oceanor in December 1996, with the due approval of the Government of India. Oceanor supplied the selected products at the agreed (lower) price, installed and maintained twelve moored data buoys in Indian waters for a year and effected the technology transfer. As planned, the NDBP installed the first of the met-ocean buoy network on 21 August 1997, off the Chennai port, using the FORV *Sagar Kanya*, ahead of the DOS launch of their remote sensing satellite OCEANSAT – the satellite, launched by ISRO for DOS.

The buoys are equipped with state-of-the-art sensors for the measurement of winds, waves, currents, atmospheric pressure, atmospheric temperature, sea surface temperature, dissolved oxygen, chlorophyl, radioactivity, hydrocarbon and turbidity. A shore station was commissioned at the NIOT for the retrieval, processing, analysis and storage of the data obtained from the buoys real-time. The met-ocean data are disseminated to the India Meteorology Department, the Indian Coast Guard and the port trusts on a daily basis. This experience subsequently helped the NIOT in indigenous developments that led to deploying tsunami buoys, an observatory in the Arctic and autonomous observation systems. The oceanographic observations from the NDBP are world-class, for which it has won awards and recognitions from UNESCO, Intergovernmental Oceanographic Commission (IOC) and the World Meteorological Organization (WMO). I extend my appreciation to K. Premkumar, programme director of the NDBP, for spearheading this programme and for his contributions towards its success.

Sea Level Observations

The NIOT is responsible for establishing tide gauge stations for the systematic and accurate monitoring of the tide levels to derive long-term variations in the mean sea level. Under the sea level observation project, the NIOT initially established ten modern tide gauge stations at Mumbai, Porbandar, Goa, Kochi, Chennai, Visakhapatnam, Paradip, Kavaratti, Machlipatnam and Thoothukudi. The data from these tide gauges were collected regularly and sent to the National Tide Data Centre at the Survey of India, Dehradun. Based on the sea-level data collected along the Indian coast, a numerical model had been constructed to simulate circulation in the northern Indian Ocean.

Drifting Buoy and Other Observing Systems

This programme, being implemented through the NIO in Goa envisages the deployment of drifting buoys, expendable bathythermographs (XBTs), and current meter arrays for the augmentation of in-situ oceanographic data. About fifty drifting buoys, three current meter arrays and 900 XBTs were deployed under this programme during the Ninth Five-year Plan period.

Remote Sensing for Ocean Observations

Many of the oceanographic phenomena, including coastal processes and climate change, are found to show variations over timescales of a few hours to years. It is therefore essential that satellite data are available consistently in the required format for a scientific understanding of the oceans and their long-term effects on human activities and the environment. Encouraged by results from several pilot-scale studies on ocean applications and the keen interest of meteorological studies in the sea-state information, the specific requirements of the country in satellite oceanography have been identified.

Towards meeting some of the urgent requirements of the user community, OCEANSAT-1, the first of the series of ocean satellites, was configured by ISRO with Ocean Colour Monitor (OCM) and Multi-Frequency Scanning Microwave Radiometer (MSMR) payloads. This satellite was launched using the indigenous PSLV-C2 on 26 May 1999, within the IRS series, to be used in ocean- and atmosphere-related applications. The OCM had eight narrow spectral bands operating in visible and near-infrared (NIR) bands (402–885 nanometres [nm]) with a spatial resolution of 360 m and swath of 1,500 km.

This OCM was meant to provide the highest spatial resolution, compared to any other contemporary satellites worldwide. It would be resourceful for the study of both physical and meteorological parameters. The OCM would provide unique regional and global observations to measure biological constituents such as chlorophyll pigments, and assess primary productivity, while the MSMR would gauge sea surface temperature, atmospheric water vapour and sea surface winds. Both of them provide the data once in two days, which could then be utilized for operational ocean applications.

ISRO had planned a series of ocean satellites in recent years to meet the critical data requirements of oceanography and ocean meteorology, assessment of primary productivity, sea surface temperature, atmospheric vapour, sea surface winds, wave spectrum, sea surface topography, ocean surface currents, internal waves and so on. ISRO thus launched OCEANSAT-2 on 23 September 2009. OCEANSAT-2 had a scatterometer, altimeter, ocean colour monitor, thermal infrared radiometer and passive microwave radiometer among other things. SCATSAT-1, the continuity mission for OCEANSAT-2 scatterometer, assists in weather forecasting, cyclone detection and tracking services and ocean state monitoring. The Satellite fort ARgos and ALtiK (SARAL) – a cooperative Indo-French altimetry technology mission to investigate the ocean meso-scale variability, monitor the main continental water level and observe polar oceans among other objectives – is a subsequent ISRO endeavour in this direction.

Ocean Information Services (OIS)

The objective of the Ocean Information Services (OIS) was to generate and disseminate user-oriented data products on an operational basis, based on in-situ and remote sensing ocean observations. Near-real-time data and data products, such as Sea Surface Temperature (SST) and Potential Fishing Zone (PFZ) advisories, upwelling zones, maps, eddies, chlorophyll and suspended sediment load, in addition to consultancy services to the Central government, states, corporate sector and industries were to be made available under one roof. Apart from ocean observations, OIS was to carry out extensive validation studies to improve the efficacy of its advisories, such as PFZ, and generate coastal wetland maps, coral reef maps and shoreline changes maps of the country, which are useful in coastal zone management to port authorities, state governments and researchers.

Ocean Dynamics and Modelling

Knowledge of ocean dynamics is a prerequisite for the prediction of the ocean state based on ocean models. The dynamics of marine ecosystems and the air–sea exchange of carbon dioxide were studied using the Modular Ocean Model (MOM). Air–sea interaction process of monsoon prediction was studied using the ocean-atmosphere coupled model. A storm surge prediction model to estimate inland inundation along the Andhra coast was developed and a seven–component nitrogen-based ecosystem model was completed.

Satellite Coastal and Oceanographic Research

The project envisages the development of algorithms and models for the retrieval of met-ocean parameters – namely, sea surface temperature, winds, waves, bathymetry, suspended matter, mixed layer depth, chlorophyll, aerosol, water vapour, clouds, currents and sea level from satellite sensors. The data assimilation models on waves and ocean circulation have been developed. The development of retrieval algorithms and validation of sea surface temperature and waves from the European Remote Sensing satellite (ERS-1) was completed, which includes the customization of the software. The software development for the estimation of coastal bathymetry from spaceborne synthetic aperture radars (SARs) and altimeters was also done. Studies on latent heat flux and eddies would be concluded from ERS and altimeter, respectively.

The Indian National Centre for Ocean Information Services (INCOIS)

As is evident from the information given so far, substantial oceanographic data was generated by different agencies for diverse applications. I observed that the integration of the ocean information and focus on the development of user-oriented data products for divergent applications, such as fisheries, navigation, weather forecast and oceanographic research, were lacking. To manage the ocean observations scenario and to generate and disseminate data products that meet the operational needs of assorted users in an effective manner, I established the Indian National Centre for Ocean Information Services (INCOIS) at Hyderabad in February 1999 as an autonomous body of the Department of Ocean Development.

The initial efforts of DOD to develop satellite oceanography were supported by the National Remote Sensing Agency (NRSA) of DOS. The activities of DOD were located in the NRSA building in Hyderabad. I approached the Government of Andhra Pradesh for separate premises for INCOIS. I am grateful to the state government for allotting 50 acres 39 *guntas* (nearly 51 acres) of land at Gajularamaram in the

Ranga Reddy district of Andhra Pradesh. On 15 December 2000, I laid the foundation stone for the new INCOIS campus. I appreciate the contributions of Prof. B.L. Deekshatulu, the director of NRSA, and thank him for his guidance and assistance. The preliminary efforts were coordinated by Dr A. Narendranath, who constantly interacted with the authorities to secure the land. The programme in DOD HQ was coordinated by B.N. Krishnamurthy and supported by Dr K. Somasundar. I appreciate their contribution to OOIS, implemented through INCOIS.

In July 2001, Dr K. Radhakrishnan from ISRO (then at NRSA) was appointed as the director of INCOIS. He established the infrastructure, developed human resources and capacity building, and evolved a programme profile for INCOIS and implemented it successfully. He transformed the centre, in a short time, into a world-class science and technology institution with a clear vision, mission and focused programmes on ocean observations, ocean modelling, satellite oceanography and ocean information services. He formed a young and dynamic team as well as instituted state-of-the-art infrastructure on the new campus. INCOIS undertook a national mission on Potential Fishing Zone (PFZ) advisories (Photo No. 72), thereby ensuring significant direct economic benefits to the fishing community along the entire coastline of India.

An international collaborative programme for better awareness of ocean circulation and enhancement of climate predictability was then evolving with the sponsorship of WMO, IOC and United Nations Environment Programme (UNEP), which required a global array of 3,000 Argo profiling floats by 2006. As part of this array, 450 Argo floats were to be deployed in the Indian Ocean, with India contributing one-third of them and the balance provided by seven other nations. In 2001, INCOIS was chosen as the regional coordinator and the Regional Argo Data Centre of this international effort for the entire Indian Ocean.

INCOIS is the prime mover in the country for modelling the ocean-atmosphere system, and it steers the implementation of projects on Indian Ocean modelling, dynamics, satellite oceanography, coastal research and so on, with the active participation of leading R&D institutions and the academia.

In the aftermath of the catastrophic tsunami of December 2004, INCOIS was chosen to set up an early warning system for tsunami and storm surges in the country, ensuring the synergy of DOD, DOS, DST and CSIR.

20
Integrated Coastal and Marine Area Management

AS CAN BE NOTED, the department established several new initiatives and paid great attention to the coastal and marine environment towards sustainable developments. These programmes were multidisciplinary and needed to be executed by multi-institutional participation. Hence, the department had to implement a number of research projects with the support of adequate oceanographic research vessels and human resources with required expertise. In addition, the department restructured and reoriented the coastal community programme both on mainland India and the Andaman and Nicobar group of islands. In this chapter four programmes of the department are presented – the first one addressing integrated coastal and marine management with focus on marine environment and coastal zone, the second one on the oceanographic research vessels including specially built coastal research vessels, the third one on capacity building in all relevant areas of science and technology and the fourth one on coastal community both on mainland India and the Andaman and Nicobar group of islands.

The sustainable development of the earth's rich and diverse coastal areas has been affected by complex patterns of human activity and various economic sectors. The coastal zone is unique, spatial and dynamic. Since the adoption of Agenda 21 at the United Nations Conference on Environment and Development (UNCED), India has been implementing the Marine Environment and Coastal Zone programme to monitor coastal waters and assist in the protection of marine environment and the sustainable development of coastal and marine areas.

The programme was redesigned in 1991 as Coastal Ocean Monitoring and Prediction System (COMAPS) to include the modelling aspects and intensive monitoring of marine environment. The COMAPS programme was planned to assess the status and likely trends in coastal marine environment quality and to contribute directly to the development of strategies for maximizing the economic benefits derived from intelligent use of marine environment and resources.

Based on the nature of ongoing coastal activities, a number of sampling stations were chosen along the coastline of India. Data on environmental (physical, chemical and biological) parameters of the water column and sediments, including pesticides, heavy metals and other pollution indicating parameters such as dissolved oxygen and pathogenic bacteria, were monitored. The data collected were reported to the Central Pollution Control Board and the coastal state pollution control boards, so as to enable them to take remedial measures.

All activities related to marine environment and coastal zone at DOD were led by Dr B.R. Subramanian. From the time when it was a small directorate at DOD, he monitored and guided the activities of its divergent research institutions and government agencies. In 1996, I reviewed all marine and coastal zone activities at DOD. Jointly with Dr B.R. Subramanian, I studied the multifarious programmes planned and executed by other nations. I recognized that Integrated Coastal and Marine Area Management (ICMAM) was a viable and proven alternative to the traditional methods of planning and management in coastal zone.

ICMAM could provide a framework and the practical tools to assist policymakers, planners and resource managers to meet the challenges of sustainable development in coastal areas. It could also help in rational utilization of marine resources and proper management of marine environment to prevent its degradation from anthropogenic activities such as pollution. When applied in a timely and comprehensive manner, ICMAM could provide a vehicle for sound investment in and sustainable use of coastal natural resources.

The department, after discussions with the Ministry of Environment and Forests, participated in the Environment Management Capacity Building project aimed to develop its capacity for environmental management of coastal zone. The project was funded by the World Bank and addressed the following areas:

- Development of GIS-based information system for critical habitats such as mangroves, coral reefs, and areas of biodiversity and turtle nesting ground, in the following eleven localities: the Sundarbans, the Gulf of Khambat, the Gulf of Kachchh, Malvan in the Sindhudurg district of Maharashtra, five islands off Karwar, three islands off Kochi, the Gulf of Mannar, Pichavaram, near Chidambaram in Tamil Nadu, Coringa mangrove in East Godavari district of Andhra Pradesh, Gahirmatha in Odisha, and Kadamat (Lakshadweep)

- Determination of waste-load allocation based on waste assimilation characteristics of selected estuaries (Tapi Estuary and Ennore Creek);
- Development of EIA guidelines for major coastal developmental activities and processes including the construction of ports, harbours and breakwaters, disposal of domestic and industrial wastes, marine transportation and coastal tourism;
- Development of model ICMAM plans; and
- Infrastructure, training, R&D and survey.

Integrated Coastal and Marine Area Management Project Directorate (ICMAM-PD)

In consultation with Dr B.R. Subramanian, and with due approvals from those concerned, I evolved R&D projects, besides the ICMAM programme, and established an independent project directorate of the department, co-located with the NIOT in Chennai. This step was to ensure an integrated approach in the management of marine environment with the following broad objectives:

- Acquiring knowledge related to advanced tools and techniques for selected activities in connection with integrated management of coastal and marine areas and development of expertise to use these tools to meet the demands of the present as well as the future

- Enhancement of the capabilities of the central and state governments, national and state-level facilities/ academic institutions and research laboratories to equip them to deal with various technical and management issues related to integrated management of coastal and marine environment through capacity-building activities, such as training and field exercises;
- Creating basic infrastructure and facilities to conduct surveys, R&D and orientation on ICMAM-related activities;
- Development capabilities in coastal states and Union Tertiaries (UT) for adoption of ICMAM through training;
- Creating a base for human resource development in the field of ICMAM; and
- Ensuring sustainable development and management of ocean resources, enforcing regulations for ocean-related activities and maintaining the Indian Ocean clean and productive.

Accordingly, an office memorandum was issued in September 1998 and an ICMAM project directorate was established in Chennai, as an independent R&D unit of DOD, with Dr B.R. Subramanian as the project director. The directorate was treated as an attached office of the department. The ICMAM project directorate, with the director, ten scientists and eight supporting staff, continued the research with full financial support of the MoES and made significant contributions in the following areas:

- Shoreline management by providing conceptual solutions to control erosion at Ennore (Tamil Nadu), from Munnambum to Kayamkulam and Panathura (Kerala), Gahirmatha and Gopalpur (Odisha), Gangavaram (Andhra), and Karwar and Mangaluru (Karnataka);
- Development of safe limits for toxic chemicals such as copper, cadmium, mercury, lead, zinc, arsenic and chlorophos;
- Modelling of the Chilika lagoon using advanced techniques and prediction of the productivity of the lagoon for fisheries, a similar prediction having been carried out for the Kochi backwaters up to secondary (zooplankton) production;
- Extensive mapping of the entire Indian coast for inundation due to tsunamis and development of modelling techniques to predict inundation;
- Modelling of coastal inundation by storm surges and validation for cyclones;
- Training programmes to enhance the knowledge of the scientists and technical staff of state governments on the application of GIS in coastal area management, more than 200 such programmes having been organized; and
- Regular interaction of the scientists of ICMAM–PD with the officials from coastal states and relevant central departments, resulting in providing a number of technical services, such as opening the inlet of Chilika lagoon, preparation of DPR for Odisha ICZM projects, and development of guidelines for erosion control.

State and UT governments extensively benefited from the products developed by ICMAM. ICMAM focused on basic and applied research, addressing scientific aspects of issues related to coast and coastal zone, thereby leading to its upgradation as the National Centre for Coastal Research (NCCR) in 2018.

NCCR aims to specifically address coastal pollution, coastal and shoreline processes, and coastal hazards with significant societal, economic and environmental benefits.

Oceanographic Research Vessels (ORVs)

Oceanographic Research Vessels (ORVs) are the backbone of oceanographic research. When I joined DOD in 1995, the department had only two of them: ORV *Sagar Kanya* (Photo No. 67), a multidisciplinary oceanographic research vessel, and the FORV *Sagar Sampada*, a multipurpose fishery oceanographic research vessel. An urgent need for coastal research vessels was felt in view of the need to collect samples in near-shore waters for coastal zone studies, especially in research on the chemistry and biological aspects of these waters. I realized that ORV *Sagar Kanya* and FORV *Sagar Sampada* could not support such studies. Hence, I decided to construct two smaller-size coastal research vessels – one each for the west and east coasts with the minimum equipment necessary.

ORV *Sagar Kanya*

In 1983, under Indo-German collaboration, a multidisciplinary research vessel was built in Germany and delivered to DOD. The vessel was a versatile ocean-observing platform equipped with technologically advanced scientific equipment and related facilities. It was built in conformity with the highest-class requirements, as specified in the Lloyd's Register of Shipping and the Indian Register of Shipping. The vessel is fully automatic diesel–electric type and equipped with a Dynamic Positioning System. In addition to the twin-screw propulsion, two fin rudders and one bow thruster give it excellent manoeuvrability.

Initially the vessel was operated and maintained by the NIO in Goa, for use in national oceanographic research projects. When I reorganized the activities of DOD, I constituted a national-level committee to guide oceanographic research programmes in association with the NIO and other academic and national laboratories. I also felt that an effective management system for maintaining the ORV *Sagar Kanya* and all its equipment was essential to promote national oceanographic research. Towards this purpose, DOD took over the management of the vessel from the NIO.

I appointed Dr M. Sudhakar, a senior scientist from the NIO, at DOD in 1997 and assigned him the responsibility of the ORV *Sagar Kanya* vessel management. I issued a contract, after following due procedure, to an external agency that had the expertise and human resources required to maintain the on-board scientific equipment. Subsequently, I transferred the ORV *Sagar Kanya* from DOD to the NCAOR and assigned the management responsibility to its director, Dr P.C. Pandey, on 30 September 1997.

ORV *Sagar Kanya* contributed to the advancement of oceanography in India in an exemplary manner, carrying out over 180 cruises till the end of my tenure in January 2001. The vessel had been the flagship of the country in all ocean-related activities for nearly two decades and was utilized for atmospheric and ocean research in both national and international programmes. The results of the studies have brought laurels to the nation.

FORV *Sagar Sampada*

The *FORV Sagar Sampada* (Photo No. 68) is a multipurpose fishery oceanographic research vessel built at the Dennebrog Shipyard Ltd, Denmark. It was commissioned on 6 November 1984. The vessel is ice-strengthened to support India's scientific programme in the Southern Ocean. It is equipped to carry out multidisciplinary research in oceanography, marine biology and fishery science. It is a platform for interdisciplinary expeditions in and around the Indian EEZ, with the participation of various institutions in the country and abroad. The state-of-the-art FORV *Sagar Sampada* has been the backbone of Indian R&D efforts on MLR, including deep-sea fishery surveys. The scientific facilities on-board are:

- Sea Bird CTD (Conductivity-Temperature-Depth) Profiler,
- Acoustic Doppler Current Profiler (75kHz),
- Skalar Autoanalyser for nutrient analysis (AutoSal),
- Multiple Plankton Net (MPN),
- Grab, Dredge,
- Bongo Net for fish egg/larvae and surface collections,
- VEL Net for fish egg/larvae collection from the upper 10 m,
- Multiple Frequency Echo Sounders (200 kHz, 120 kHz and 38 kHz),

- Side Scan Sonar,
- Integrated Trawl Instrumentation System,
- Integrated Fish Finding System,
- Automatic Weather Stations (AWS),
- Facilities for fishing operations up to a depth of 1,200m, and
- Facilities for sediment core sample collection.

The vessel can accommodate twenty-four research personnel and thirty-five crew members. It is equipped with Fleet 77, a new maritime satellite service, for communication, satellite navigator for navigation, helipad for the emergency landing of a helicopter, and a medical room with a qualified doctor. The FORV *Sagar Sampada* has an average speed of 10 knots and an endurance of thirty-five days.

Since DOD did not possess adequate infrastructure and human resources to manage this vessel, the CMFRI of ICAR was given the responsibility of its maintenance and operation. The vessel provided facilities and services to a number of fisheries research institutes, such as the Fishery Survey of India (FSI), the Central Marine Fisheries Research Institute (CMFRI), the Central Institute of Fisheries Technology (CIFT), the National Institute of Oceanography (NIO), and many other academic institutions. I noticed that it would need to be improved to be able to meet and keep up with the requirements of these diverse institutions. I therefore established a Sagar Sampada Cell in Kochi as an attached office of DOD, which would maintain and operate the vessel, and placed in position a system and procedure to plan and execute the cruises. I also gave a contract to an external agency to service and maintain all the scientific equipment on-board.

The vessel was mainly used for surveys in India's EEZ. During the Ninth Five-Year Plan period (1997–2002), it was fully dedicated to studies on the MLR of the Indian Ocean. Research on the environment, productivity, fishery resources, deep scattering layers, benthic production and toxic algae in the Indian EEZ were carried out through fifty-six dedicated cruises.

After necessary modifications, I employed this vessel for the first time to survey fishery resources in the Antarctic waters. It was during the winter of 1995–96. At present, the FORV *Sagar Sampada* is managed by the upgraded Centre for Marine Living Resources and Ecology (CMLRE), an important R&D centre of DOD.

Coastal Research Vessels (CRVs) Sagar Purvi and Sagar Paschimi

Coastal zone studies, especially research on chemical and biological aspects, require extensive collection of samples from the sea. Generally, small fishing vessels are used for the purpose in near-shore waters, while large ocean-going vessels provide facilities for the collection of deep-sea samples. As the near-shore coastal waters up to 50 m depth are zones of intensive physical, chemical and biological activities, it is extremely essential to conduct investigations in this area. While large vessels can navigate up to a minimum 20 m of depth, they are of limited use for shallow water research. Fishing boats have limited endurance of a day or two, which is far from adequate for the continuous monitoring of chemical parameters in the sea.

67. Oceanographic Research Vessel (ORV) *Sagar Kanya*

68. Fisheries Oceanographic Research Vessel (FORV) *Sagar Sampada*

69. Coastal Research Vessel (CRV) *Sagar Purvi*

70. Coastal Research Vessel (CRV) *Sagar Paschimi*

71. Culture of mud crab in cages in the Andaman and Nicobar Islands

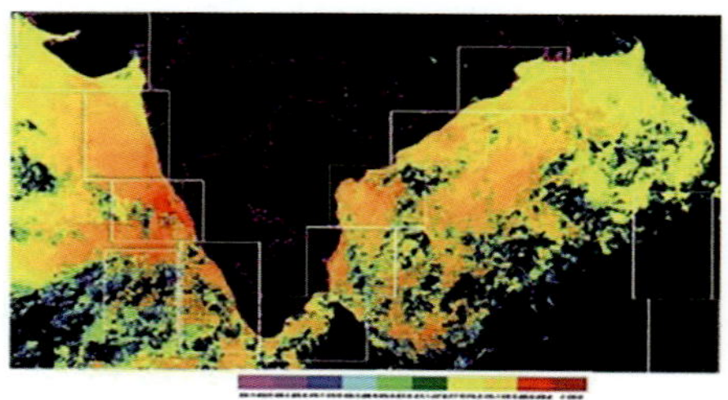

72. Sea surface temperature composite (recorded in December 2002) to generate potential fishing zone (PFZ) advisories

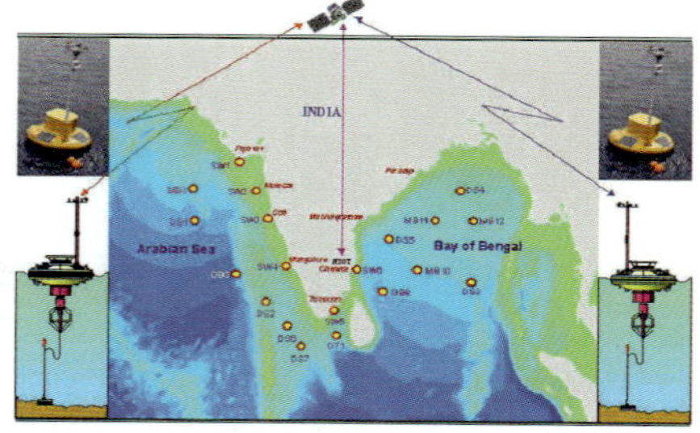

73. National data buoy programme (NDBP) of NIOT, deploying a 20-buoy network in Indian seas

74 Delineation of the outer limits of the continental shelf

Faced with the task of having to meet these requirements, DOD considered the design and fabrication of two coastal research vessels – one each for the east and west coasts of India. Dr B.R. Subramanian played a major role in organizing and implementing the project. The two vessels were designed indigenously in 1994 and were the first oceanographic vessels built in India. I closely monitored the construction of these two vessels, held review meetings with the consultants – S. Paramanandam, Director General of the Naval Design Directorate, and Dr P. Sambandan, the director of National Ship Design and Research Centre (NSDRC) and ensured the timely completion of the work.

The vessels were named *Sagar Purvi* (Photo No. 69) – to operate along the east coast, built at Corporate Consultants, a company in the private sector in Kolkata – and *Sagar Paschimi*, (Photo No. 70) to be deployed along the west coast and built at Alcock Ashdown, a company in the private sector in Bhavnagar. I inaugurated both of them in December 1996. Along with Dr B.R. Subramanian and Professor M. Ravindran, the former director of the NIOT, I established a Vessel Management Cell (VMC) with six dedicated technical staff. Standard operating procedures to operate and maintain the vessels were formulated in consultation with experts. The vessels, though meant for normal weather, plied during moderately adverse weather conditions and provided extensive services to the coastal oceanographic research community. The number of average vessel days covered in a year was 180–200, against the stipulated 100–150 days.

The vessels had sampling gears, such as special winches to collect seawater and sediment and fishing winches to collect fish samples. The water quality laboratory was equipped with sample treatment equipment and analytical instruments such as spectrophotometer. It had cold storage facilities for the preservation of samples. The navigation system was also sophisticated, with satellite communication facilities. Both vessels provided extensive support in coastal research, especially with pollution monitoring which led to clear identification of zones of low, moderate and high pollution in the coastal waters of India. It helped in extensive documentation on the biodiversity of coastal waters and their current status of abundance.

The vessels played a vital role in conducting surveys related to the establishment of desalination plants in Lakshadweep Islands. While *Sagar Paschimi* was decommissioned in October 2014 after it was severely damaged in Cyclone Hudhud that crossed the Visakhapatnam coast that year, *Sagar Purvi* continued to assist the research community. The knowledge and experience acquired in the design and construction of the two vessels led to indigenous expertise in this area.

Capacity Building in the Ocean Sector

India is represented in almost all the important international and intergovernmental bodies on oceans. The activities in the ocean sector are national, regional and global, with scientific, economic and geopolitical impacts. The country needs an integrated approach for the scientific planning, formulation, implementation and overall management of the programmes with comprehensive organizational and institutional backup. The perspective ought to address scientific research and technology development with an orientation towards applications, obligations and privileges arising out of international conventions and treaties, and bilateral and multilateral cooperative ventures.

DOD recognized that an essential requirement towards these concerns was capacity building and human resource development for self-reliance. Hence, from the very beginning, the department sponsored research programmes in a wide spectrum of the ocean sector, initially focusing on ocean sciences. Dr V.S. Chinthala Rao was nominated coordinator for capacity building in the department. Dr S. Varadarajan, former director general of CSIR, had been guiding the department as an honorary chief executive of the capacity-building programme and administering it through the coordinator. The department received research proposals and processed them for approval, modification or rejection after consulting Dr Varadarajan who was an expert. I recognized that Dr Varadarajan had critically reviewed a large number of proposals and rendered valuable suggestions to implement the capacity-building programme across the country. I appreciate his excellent work and thank him for his contributions.

I observed that a large number of proposals were administered centrally but a system to carry forward the research findings towards applications for the benefit of society and national development was lacking. I also felt that in order to create specialized manpower in the ocean sector, there was a need to focus on the newly emerging frontier areas of coastal engineering, marine chemistry, marine biology, marine ecology, marine microbiology, pharmacology, coastal zone management, marine archaeology, exploration and exploitation of marine resources and so on. Among the proposals for capacity building, the majority were from academic institutions. Over a period of time, the department established a good working relationship with several universities. As a part of the efforts to promote research and stimulate original, path-breaking scientific work on oceans, I realized the need to support such activities organized by universities and national laboratories, geared towards strengthening of science infrastructure and human resource pool in ocean science and technology.

In order to create a platform to generate relevant technologies, I decentralized the capacity-building programme by establishing Ocean Science and Technology Cells (OSTCs) in different coastal universities for the promotion of front-ranking research in key focus areas in specified fields. Thus, the OSTCs were associated with many universities that had expertise in specific fields. Each OSTC was governed by a management board under the chairmanship of the vice chancellor of the particular university. The directors of autonomous institutes under DOD (Dr P.C. Pandey from NCOAR and Dr M. Ravindran from NIOT) and the programme directors at DOD (Dr V.S. Chinthala Rao, Dr B.R. Subramanian, Dr V. Sampath and V. Ravindranathan) became the vice chairmen. Professors with expertise in the specified field, from within and outside the university, were also included as members. The head of the department in the university with the subject expertise required was nominated as member secretary.

The human resources and infrastructure facilities available in different universities in the disciplines concerned were brought together. The department awarded special grants to the OSTCs for augmenting the research facilities and making them available to users from the university and outside. Proposals to the department for financial support were first referred to the OSTC concerned and only acted upon by the department thereafter, as per the recommendations of the respective management boards of the OSTCs.

The OSTCs monitored the progress of the research activities, reviewed and guided them and sent periodical reports to the department with recommendations. I also constituted a steering committee under my chairmanship, with all the chairmen, vice chairmen and research coordinators of the various OSTCs as members and the programme director in DOD as member convenor. This committee periodically reviewed the performance of these OSTCs, and resolved any issues in the implementation of the programme. It also approved the recommendations made by the management boards with modifications, if necessary. Nine OSTCs were established during my tenure at DOD. They were:

1. Ocean Engineering and Underwater Robotics cell at IIT Kharagpur in 1999,
2. Marine Microbiology cell at Goa University in 1998,
3. Marine Geology and Geophysics cell at Mangalore University in 1998,
4. Beach Placers cell at Tamil University in 1999,
5. Coastal Marine Culture Systems cell at Andhra University in 1998,
6. Marine Biology cell at Annamalai University in 1999,
7. Marine Benthos cell at the Cochin University of Science and Technology in 1999,
8. Marine Coastal Ecology of West Coast cell at Bhavnagar University in 1997, and
9. Marine Coastal Ecology of East Coast cell at Berhampur University in 1998.

More than a hundred projects were sanctioned to these OSTCs under various disciplines for the promotion of basic and applied research and capacity building in the ocean sector. I had visited all these OSTCs, seen their facilities, held discussions with the vice chancellor/director, registrar, and the professors concerned and requested their support for the capacity-building programme of the department in the ocean sector. In addition, a large number of fellowships were awarded to young scientists through these OSTCs to augment the manpower, assist in capacity building and improve the scope of further research in marine sciences.

During the Tenth Five-Year Plan, the Government of India constituted the Ministry of Earth Sciences (MoES) by integrating DOD and the atmospheric sciences and seismology-related activities of the Department of Science and Technology (DST). Consequently, OSTCs were renamed Ocean and Atmospheric Science and Technology Cells (OASTCs). In 2006, based on their excellent performance, the Marine Biology cell at Annamalai University and Marine Microbiology cell at Goa University were upgraded to Centres of Excellence (CoEs). I understand that these two OSTCs have made significant contributions to the disciplines concerned under the able leadership of their respective research coordinators – Dr T. Balasubramanian of Annamalai University, and Dr Saroj Bhosle of Goa University. Dr V. Sampath, former adviser of the MoES and vice-chairman of these OSTCs, too worked closely with them and paved the way for the promotion of these institutions into CoEs. I congratulate all of them for their contributions in getting these two cells elevated to Centres of Excellence.

In November 2009, the secretary to the MoES constituted a committee under my chairmanship to review the performance of the OASTCs and submit recommendations. The committee consisted of GOI's Controller General of Accounts Aruna Makhan, former NIO scientist Dr Mohideen Wafar, former deputy director general of the Geological Survey of India (GSI) Dr B.K. Saha, and former director of the Centre for Earth Science Studies (CESS) Dr M. Baba. Dr P.G. Dastidar was the OASTC programme officer. The committee visited all the OSTCs and held discussions with the vice chancellors, members of the management board and professors concerned. The committee also oversaw the facilities established over the years for the OSTC programmes. Presentations were made about the status and progress of the capacity-building programme, with special reference to interactions with and pooling the resources of other universities. The committee had separate internal meetings and generated an objective evaluation report. This report was submitted to the secretary, MoES, by the member secretary in October 2010 for consideration and further action.

Community Programmes in the Ocean Sector

Between 1995 and 2001, I restructured and reoriented the coastal community programmes, both on mainland India and the Andaman and Nicobar group of islands. The activities included:

- Phase-II of the shore to fishing vessel communication system to ensure the safety of fishermen at sea,
- Design and fabrication of an integrated GPS-cum-fish finder to easily locate the fish shoals and the fishing craft in the sea, and

- Island development programme that encompassed an assortment of schemes that employed technology for the controlled breeding and rearing of lobsters, commercially important finfishes, mud crab fattening, coral reef monitoring, and pollution monitoring and abatement, to name a few.

At the DOD headquarters, I was ably supported by Dr V. Sampath and his team in implementing these programmes. The demonstration of the efficacy of shore to fishing vessel communication has prompted coastal states such as Odisha, Andhra Pradesh and Tamil Nadu to set up a two-way communication system. They have established shore stations with facilities to transmit weather, cyclone and tsunami warning through VHF (Very High Frequency), HF (High Frequency), DAT (Distress Alert Transmitter) and, of late, satellite phones as well.

As recommended by the Inter-Departmental Steering Group on Ocean Instruments and Systems, the work on the design and development of various fish finders for both vertical and horizontal scanning was sponsored by the Department of Electronics and taken up at the Electronics Research and Development Centre (ER&DC), Thiruvananthapuram. Based on the feedback from fisherfolk, I understand that the technology has been commercialized for the production of integrated GPS-cum-fish-finders, which are at present used for faster fishing at sea. Fishermen follow the PFZ forecasts provided by INCOIS and save fuel, which is the major input in marine fishing operations, thereby reducing the operational expenses.

Shore to Fishing Vessel Communication

Traditional craft and medium-sized fishing vessels operating in the coastal waters of the country are exposed to high risks and have no means of communication with the shore. Realizing the need for such a communication system, the department had taken up a project to demonstrate the technology to the fishing community through a shore to fishing vessel communication system. In Phase I of the project, initiated during 1991–92, the facilities were commissioned during 1993–94 at Machilipatnam and Kalingapatnam (Andhra Pradesh), Malpe and Honnavar (Karnataka), Thoothukudi and Nagapattinam (Tamil Nadu), Neendakara and Koyilandi (Kerala), and South Goa. Nearly 850 walkie-talkie sets were distributed to small-scale fishermen to demonstrate the advantages of the system to the fishing community.

Phase II of the shore to fishing vessel communication system was implemented during my tenure at DOD, in the Ninth Five-Year Plan period. Technical specifications for the communication system were revised on the basis of the experience gained while implementing Phase-I, by involving the services of Dr D. Srinivasan, the former director of NPOL (Naval Physical and Oceanographic Laboratory, DRDO, Kochi) and the scientists of the Department of Electronics, Government of India. The height of the tower was raised from about 30 m to 40 m to increase the range of coverage to approximately 40 km from the shore for both reception and transmission. Specifications for shore transceiver and the walkie-talkies were revised for providing a minimum coverage of 40 km. The other modifications incorporated included the provision of a waterproof case for the walkie-talkies (Marine grade), single unit battery chargers instead of multi-unit chargers, provision of spare nickel–cadmium (Ni-Cd) batteries, and so on. The scheme, in Phase II, was extended to ten locations:

1. Shankarpur and Fraserganj in West Bengal,
2. Paradeep and Chandipur in Odisha,
3. Sasoon Dock and Mirkarwada in Maharashtra,
4. Saligao in North Goa,
5. Marine Hill, Port Blair, in Andaman and Nicobar Islands,
6. Periya Veerampattinam in Puducherry, and
7. Minicoy in Lakshadweep.

The task of identifying a suitable firm for the implementation of this programme was completed and the job was awarded to Punjab Wireless Systems Ltd, Chandigarh. The Society for Applied Microwave Electronics Engineering and Research (SAMEER), under the Department of Electronics, was chosen as the consultancy agency for providing the requisite services to DOD. These services included site survey, inspection of the equipment proposed to be supplied by the Punjab-based company, quality control, testing and installation and commissioning of the systems at the aforesaid places, construction of the shore station building, erection of the tower, supply of walkie-talkies and making them operational. The construction of the shore stations, erection of the tower/mast, installation of the shore station equipment, and distribution of the walkie-talkie sets were completed and the systems were commissioned in phases during 1998–99 (six locations) and 1999–2000 (four locations).

Integrated Fish Finder and Navigation Systems

During 1995–96, an order was placed with the ER&DC for the fabrication and supply of 100 units of Integrated Fish Finder and Navigational Global Positioning System (IFFNGS) for distribution to small-scale mechanized fishers. During 1996–97, ER&DC transferred the fish finder technology to two Bengaluru-based companies, Aerospace Ltd and Bharat Electronics Ltd (BEL). A test unit fabricated by these organizations was tested for environmental aspects using a fishing vessel of CIFT, Kochi.

Based on the positive results, fifty units of IFFNGS were made by BEL in 1997–98 and made available to fishermen selected by the states of Tamil Nadu, Andhra Pradesh, Karnataka, Kerala and Goa, in order to demonstrate the efficiency of these systems in locating fishing shoals and the position of fishing vessels at sea. During 1998–99, fifty more units were distributed to fishermen in West Bengal, Odisha, Gujarat, Maharashtra, Puducherry, and Andaman and Nicobar Islands. The success of the technology application was evident from the positive feedback received from the users.

Island Development Programme

The project on the demonstration of shrimp culture in the island ecosystem, implemented by DOD through the Marine Products Export Development Authority (MPEDA) from 1990–91, concluded in 1995. It imparted the standardization of technology for the development of brackish-water aquaculture in acid sulphate soil in the Andaman and Nicobar Islands, the economic viability of shrimp farming, and training to the Islanders. The MPEDA executed this programme through an autonomous society – the Andaman and Nicobar Shrimp Farming Development and Research Agency (A&NSFDRA).

On completion of this project, this society was dissolved by the MPEDA with the approval of the Ministry of Commerce and a decision was taken to transfer the facilities and assets to the Andaman and Nicobar Administration. At that point in time, the options to expand DOD's programmes in the Andaman and Nicobar Islands was explored and it was felt advantageous for the department to take over these assets and liabilities. On the department's request, the original decision was reversed and DOD took over the facilities at Port Blair from the MPEDA in February 1998. This move paid rich dividends in expanding the department's programmes on the islands for the benefit of the inhabitants.

The department had been supporting a multidisciplinary project executed by various institutions in order to have a good knowledge on the resources, ecology and environmental aspects of the coastal and marine ecosystem of the Andaman and Nicobar Islands. To coordinate these efforts and to provide working and instrumental facilities for island-based oceanographic studies, the Andaman and Nicobar Centre for Ocean Development (ANCOD) was set up at Port Blair. During my tenure, ANCOD was adequately strengthened for ongoing programmes such as monitoring the coastal ocean pollution in the Andaman waters and coral reef rejuvenation. The coordination of new programmes proposed, such as the enhancement of the marine living resources through lobster and mud crab fattening, and marine ornamental fish culture were also undertaken.

On the basis of the recommendations of the Island Development Authority, the need for a carefully thought-out science and technology development programme in the ocean sector for the islands was introduced during the Ninth Five-Year Plan (1997–2002), which involved budgetary support of Rs 4.00 crore. The programme envisaged giving impetus to achieve various charters of the department in the field for the demonstration of technology that had a bearing on the socio-economic aspects of the island community. The new schemes were:

- Enhancement of marine living resources with particular reference to technologies that augmented the stock of commercial finfish and shellfish species, marine ornamental fishes, seaweed, etc.,
- Introduction of physical oceanographic studies in the island groups as part of the ongoing programme on pollution monitoring under the COMAPS programme, and
- Setting up of tide gauges to measure tidal fluctuations and sea level rise.

In addition, programmes on the rejuvenation of coral reefs, monitoring of pollution in different transects already identified in and around Port Blair, laboratory facilities for research in various disciplines of marine science and so on continued.

In March 1998, the department approved the project for Living Marine Resources Enhancement (LMRE) through lobster rearing and fattening at a total cost of Rs 133.24 lakh, for implementation through the NIOT. The project entailed the participation of the Tamil Nadu government's Department of Fisheries, the Andaman and Nicobar Administration's Department of Fisheries, the University of Madras's Department of Zoology, and the Aquaculture Foundation of India, Chennai. The activities of ANCOD were also integrated with the NIOT to form the Andaman and Nicobar Centre for Ocean Science and Technology (ANCOST) in order to implement the core activities previously initiated under the LMRE programme.

To give a head start to the activities, a seafront laboratory and a marine animal grow-out system were developed near Chennai and various experimental programmes related to the development of fattening and larval rearing technology were taken up. The support provided by Dr M. Sakthivel of the Aquaculture Foundation of India and Dr V. Sampath, the director of DOD then, in this initiative is worth mentioning.

The centre is now known as the Atal Centre for Ocean Science and Technology for Islands (ACOSTI). An upgraded laboratory has also been set up at ACOSTI, Port Blair, to carry out studies on marine living resources, coastal environment, ocean observation and related areas in ocean science and technology.

The DOD programmes aimed at coastal development enhanced the daily life of our citizens, augmented their safety at work and contributed to their livelihood. All the activities that I had coordinated during my tenure at DOD and the resources I had identified at various levels primarily sought to apply updated technology for the benefit of the nation. It was with immense satisfaction that I stepped down from the position of the secretary to the Government of India on 31 January 2001.

21
Ocean Technologies at NIOT

THE DEPARTMENT OF OCEAN Development (DOD) established the National Institute of Ocean Technology (NIOT) in November 1993 as the first R&D establishment of the department inside the IIT Madras campus. Professor M. Ravindran of IIT Madras was appointed as the executive director of the NIOT in November 1993. Director of IIT Madras was the first chairman of the governing council of NIOT. Consequent upon the decision that secretary, DOD, shall be the ex officio chairman of the governing council, I worked as its chairman during my tenure from April 1995 till January 2001. I decided that the NIOT should take up pioneering technology development and technology demonstration projects, in addition to important 'operational programmes' on behalf of DOD.

This thought prompted me to develop the NIOT as a Centre of Excellence in ocean technology and support the ocean technology programmes of the department. In 1999, I established nine Ocean Science and Technology Cells (OSTCs) in coastal universities and academic institutions. The NIOT was identified with the OSTC on Ocean Engineering and Underwater Robotics, IIT Kharagpur.

Over the years, I guided the NIOT as it implemented a number of multi-institutional and multidisciplinary programmes on ocean technology in our country. Towards this objective, a much larger space was required for offices, laboratories, infrastructure facilities and so on. The then chief secretary of the Tamil Nadu government, with whom I enjoyed a cordial relationship, acceded to my request to transfer 50 acres of government land at Pallikaranai at a nominal price to the NIOT. The property was allotted in June 1997. I evolved a master plan and laid the foundation stone for the development on 21 August 1997.

The NIOT established the requisite facilities and began functioning with activities in the Integration Bay in December 1997. The fully operational NIOT main building, which was completed in a record time of eighteen months, was formally inaugurated by Dr Murli Manohar Joshi, the Union minister in charge of DOD, in September 2000. The efforts of Professor M. Ravindran and Dr V. Rajendran in making the campus fully operational, within eighteen months when funds were sanctioned in March 1998, are well appreciated.

The NIOT implemented several projects on ocean technology during my tenure. This included the National Data Buoy Programme (NDBP) of 1993, already presented in the chapter on OOIS. Real-time data, from various met-ocean sensors, received eight times daily at three-hourly intervals via satellite, are transmitted after the necessary quality control to user agencies such as the India Meteorological Department (IMD), coastguard, navy and ports. The data from these buoys have become an indispensable part of the weather forecasting programmes of the IMD. Special buoys have been installed under the NDBP for the Bay of Bengal Monsoon Experiment (BOBMEX) and Arabian Sea Monsoon Experiment (ARMEX) cruises organized by DST to monitor air–sea interactions in the Bay of Bengal and the Arabian Sea during the monsoons. The contributions of NIOT to the ocean mining technology development under the already reported marine non-living resources programme is praiseworthy.

Coastal Research and Management

I involved the NIOT in technical consultancy projects, mainly in the area of coastal and environmental engineering. These project sites, numbering more than a hundred, were spread along the entire coast of India, including the Andaman and Nicobar Islands and Lakshadweep. Each project, being different in nature and having unique site conditions, contained a new research component to a certain extent, apart from giving valuable technical outputs. Some of the sophisticated offshore survey equipment and software in the area of physical oceanography, seismography and geophysics and multibeam bathymetry were handled by this group in India.

In December 1996, to promote research on coastal and shoreline processes, DOD launched two coastal research vessels *Sagar Purvi* and *Sagar Paschim*i – *w*ith on-board oceanographic instruments and laboratory for the Coastal Ocean Monitoring and Prediction System (COMAPS) programme. It contributed to the national capacity building on operational oceanography, promoted inter-institutional cooperation and research projects, and aided the collection of vast volumes of oceanographic data. The institute was vested with the responsibility of managing these vessels by setting up a Vessel Management Cell (VMC) during 1996–97.

Marine Instrumentation

The marine instrumentation group of this institute concentrated on the development of acoustics-based instruments, since acoustic systems are the best suited for underwater applications. The first product of this group, the Acoustic Tide Gauge, has been awarded a US patent for its unique self-calibrating system. Such advanced tide gauges have been installed at six ports in India, and at Qui Nhou, Vietnam. This group has also developed a wide range of acoustic transducers that meet divergent applications in different frequency ranges.

Marine Living Resources Enhancement

In 1998, I initiated a programme on marine living resources enhancement at the NIOT under the Island Development Programme. In this venture, breeding, larval rearing, fattening and sea ranching of the Indian spiny lobster and mud crab fattening (Photo No. 71) were taken up in various parts of the country, including the Andaman Islands. After highly successful demonstrations, the programme was directed towards socio- economic development. Adopting the open sea cage culture of marine fin fishes in the coastal areas of the country, it engaged in marine micro-algal culture for nutraceuticals and biofuel, deep-sea microbial culture, the deployment of artificial reefs in Odisha and fish aggregating devices in the Andaman and Lakshadweep group of islands among others.

Gas Hydrates

In the International Long-Term Programme (ILTP) meeting at Delhi in 2000, I proposed that the responsibility of the Indian gas hydrate programme be assigned to DOD. This idea, which had been under discussion for a long time, was accepted. My concerted efforts with the Russian Academy of Sciences (RAS) and the follow-up on the part of my successor led to an MOU signed by the secretary to the DST and the president of RAS, in November 2003, and the establishment of the Indo-Russian Centre on Gas Hydrate Studies (IRCGHS) at the NIOT in February 2004. The Ministry of Earth Sciences (erstwhile DOD) of India entrusted the project on 'Gas hydrate exploration and technology development for its exploitation' to the NIOT to be implemented jointly with the National Geophysical Research Institute (NGRI) and NIO as an ongoing venture.

Ocean Thermal Energy

In 1992, the Tamil Nadu government had entered into an MOU with the Sea Solar Corporation of USA to establish a 100 MW Ocean Thermal Energy Conversion (OTEC) plant. Recognizing the lack of progress and the need for technology transfer, I discussed the possibility of a collaboration with Professor T. Uehara of Saga University in Japan. Based on successful negotiations, an MOU was signed between the NIOT and Saga University on 30 September 1997, for technology development and culmination of an optimized commercial OTEC plant in India.

In June 1998, the Government of India approved a pilot OTEC project with 1 MW capacity (gross). The NIOT carried out the development in close collaboration with Saga University. With the active support of DOD, a special flat-bottomed barge was built by the Goa Shipyard, with NSDRC assistance. I personally monitored the progress and held periodic reviews with the NSDRC and the Goa Shipyard. However, the developments of all the subsystems could not be completed during my tenure. I understand that the NIOT developed some of the subsystems successfully, but the demonstration of the integrated OTEC system failed in two consecutive field tests. The project design and the budget were revised twice, with the final budget rising to about twice the original value and the final project duration was nearly thrice the original timeline. Since the finance ministry did not approve of further revision, the project was short-closed in 2004 without demonstrating even a single integrated system test.

Desalination

The NIOT initiated technology developments in the area of desalination after my tenure. On retiring, in early 2001, I finally returned to academics. I joined Karunya University in Tamil Nadu and worked towards the development of an economical, environment-friendly and reliable desalination system, different from the operational systems until then. My attempts were supported by the DST, under the Science and Engineering Research Council (SERC) scheme, through two research projects, one in 2002 and the other in 2004. Soon,

I received an Indian patent (No. 196396 of 2003), for Low Pressure Distillation (LPD) for desalination.

This LPD process was successfully demonstrated by Karunya University, in association with two private industries, under my leadership in an experimental desalination plant, of 100 litres per hour capacity, at the Thoothukudi Thermal Power Station in October 2004. The desalination plant used the warm saline water discharge from the thermal plant to the sea as feed water and the cold saline water drawn from the sea as cooling water. The power required for the operation of pumps to circulate the warm and cold saline waters and to maintain the vacuum in the system was supplied by the thermal station. The Low Temperature Thermal Desalination (LTTD) plants of the NIOT, set up at Kavaratti in 2005, adopted the same patented process for desalination, incorporating the engineering and configuration required to suit the capacity and the plant site.

Desalination Proposal for Consideration

I find great potential in an Integrated Desalination System (IDS), wherein LPD is integrated with OTEC, and OTEC power is used along with its warm and cold saline discharges to the ocean. This IDS is environment-friendly and sustainable, and does not require any external power supply for operation. In this context, the Makai Ocean Engineering Company, the world leader in OTEC technology development, has expressed its willingness to be the technology provider for the establishment of an OTEC-plus-LPD desalination plant jointly with a consortium of Indian industries and academic institutions. I discussed a collaborative project on desalination with the secretary, MoES, and submitted a proposal. I have received no response so far.

A Memorable Event

I planned to carry out the preliminary field demonstration of the shallow bed mining crawler, developed at a cost of around Rs 8 crore, jointly by the IKS, University of Siegen and the NIOT off the Goa coast. The ORV *Sagar Kanya* was deployed for the purpose. Dr P.C. Pandey, the director of NCAOR, and Dr M. Sudhakar, in-charge of the ORV *Sagar Kanya* vessel management, coordinated with the NIOT and executed the field trials. After the preliminary field trials, the crawler was lifted to be placed on-board *Sagar Kanya*. As the crawler was being lifted up, through the A-Frame of *Sagar Kanya*, the lifting wire unfortunately snapped and the crawler went down into the ocean bed. The disappointed scientists returned to the port late in the evening and informed me of this setback. I held discussions with Dr P.C. Pandey, Dr M. Sudhakar, Prof. M. Ravindran and other team members from the NIOT regarding the various options. When we contacted professionals with experience in sea recovery operations, the payment demanded was rather high to the tune of a few crore rupees.

The battery in the equipment of the crawler was operational and signals were received. Thus, we had one option: locate the crawler, reach it with the requisite facilities and pull it up with the assistance of divers. I held detailed discussions with Dr M. Sudhakar and finalized an action plan. A decision had to be made without delay, since the batteries in the crawler would die after a few hours which would make it even more difficult to locate. I decided to make use of *Sagar Kanya* for the 'recovery operations' under the command of Dr M. Sudhakar, exploit the services of professional divers and start the procedure early the following day. The next morning around 5 a.m., I issued instructions to Dr Pandey.

Dr M. Sudhakar made all the arrangements overnight and *Sagar Kanya* sailed to recover the crawler. Focusing on the signal received from the submerged crawler, we managed to locate its position. Fortunately, the depth of the seabed on which the crawler was lying was around 30 m. The divers located the crawler, tied ropes to it, pulled it up with the help of *Sagar Kanya* and brought it safely on-board. It was an occasion of great joy. Immediately upon *Sagar Kanya*'s triumphant arrival at the Goa port, I conveyed my appreciation and thanks through Dr Pandey to Dr Sudhakar, Dr Ravindran and the members of the team that had successfully carried out the operation. Dr Pandey's timely response and leadership saved the day.

22

International Cooperation in the Ocean Sector

THE UNITED NATIONS CONVENTION on the Law of the Sea (UNCLOS) – also called the Law of the Sea Convention or the Law of the Sea Treaty – is the international agreement that resulted from the third United Nations Conference on the Law of the Sea (UNCLOS III), which took place between 1973 and 1982. It defines the rights and responsibilities of nations with respect to their use of the world's oceans, and formulates guidelines for businesses, the environment and the management of marine natural resources. The Convention was formally opened for signature in Jamaica on 10 December 1982, and came into force on 16 November 1994. India ratified UNCLOS in June 1995.

While the UN Secretary General receives instruments of ratification and accession and the UN provides support for meetings of states that are party to the Convention, the UN has no direct operational role in its implementation.

Organizations such as the International Maritime Organization (IMO) and the International Seabed Authority (ISA), which were established by the UN Convention and the International Whaling Commission (IWC), however, play a role. ISA is an intergovernmental body based in Kingston, Jamaica, that was formed to organize, regulate and control all mineral-related activities in the international seabed area beyond the limits of national jurisdiction, a region underlying most of the world's oceans.

The Indian Continental Shelf Programme

Article 76 of UNCLOS provides every coastal country jurisdiction over its Exclusive Economic Zone (EEZ), the region up to 200 nautical miles (M) from the baseline (i.e., the defined tide line on shore), where it may exercise sovereign rights to explore and exploit living and non-living resources of the seabed and the waters. Further, it extends dominion over an 'extended continental shelf' beyond the 200 M limit, over which the state may exercise authority over the living and non-living resources of the seabed and the sub-seabed; but not over the overlying waters. Every coastal state desirous of establishing its outer limits, if its continental shelf extends beyond the 200 M limit, should document the process of delineating its outer continental shelf within a prescribed time limit. This information must then be submitted to the UN Commission on the Limits of the Continental Shelf (CLCS) in pursuance of Article 76 of UNCLOS as well as the Scientific and Technical Guidelines of CLCS.

I noticed that the continental shelf of India extended beyond its 200 M limit and felt that a major national level project was essential to assert our national sovereign rights over the shelf beyond EEZ. In consultation with the Ministry of External Affairs, Ministry of Defence, and Ministry of Science and Technology, I initiated the Indian Continental Shelf Project in 1999. It was a multi-institutional, mission-mode project in association with the NCAOR, NGRI, NIO, DGH, NHO and Indian Navy, among others, with a mandate to generate the requisite scientific and technical data for submission to CLCS as per its guidelines. The underpinning efforts were comprehensive including multi-institutional marine geophysical surveys and data analysis that comprised state-of-the-art multichannel seismic reflection, refraction, gravity and magnetic data profiling, in addition to bathymetric surveys.

The action plans were finalized and the project was initiated in DOD, which entrusted the responsibility to the NCAOR. Geo-scientific data of an offshore area of over 1.2 million sq. km had to be generated, analysed, interpreted and documented (Photo No. 74) as per CLCS guidelines, and submitted on time. I understand that it was implemented as a flagship project of DOD and that the documents were handed over to the CLCS in May 2009. The entire data set has been archived at the NCAOR with restricted access to outsiders, in view of the sensitivity involved. I congratulate Dr P.C. Pandey, Dr S. Rajan, Dr M.P. Wakdikar and their team for the magnificent work of national importance.

The Intergovernmental Oceanographic Commission (IOC)

The Intergovernmental Oceanographic Commission (IOC) was established in 1960 as a UNESCO body to promote marine scientific investigations and ensure cooperation among the member states regarding ocean research and services. India is a member of the IOC assembly and its executive council. I was one of the four vice chairmen of IOC during 1996–98. Officials from India serve as experts on various committees of IOC.

Regional Workshop on Capacity Building

During 18–19 November 1996, a regional workshop on capacity building was convened by the Director General of UNESCO and hosted by the Government of India through DOD. It was organized by the NIO in Goa. The objective of this workshop was to discuss and assess capacity-building requirements, such as infrastructure, equipment and training of personnel in the member countries of the Intergovernmental Oceanographic Commission for the Central Indian Ocean (IOCINDIO) region. The workshop prioritized the Global Ocean Observing System (GOOS) modules in the Indian Ocean region and identified areas of interest for regional cooperation. Officials of DOD actively participated in this regional workshop.

IOCINDIO

India is a member country of the regional committee of IOCINDIO. The other members are Oman, the United Arab Emirates, Qatar, Saudi Arabia, Kuwait, Iraq, the Islamic Republic of Iran, Pakistan, Sri Lanka, Maldives, Bangladesh, Myanmar, Thailand, Malaysia, Indonesia, Australia, the United Kingdom and France. During 20–22 November 1996, the second session of the regional committee of IOCINDIO

(IOCINDIO-II) was held at the Cidade de Goa Hotel in Goa to promote cooperative oceanographic research and services in the waters of the Central Indian Ocean. Among the member countries of the regional committee of IOCINDIO, Qatar, the Islamic Republic of Iran, Sri Lanka, Maldives, Bangladesh, Indonesia, Australia and India participated in the meeting.

On 20 November 1996, Dr E. Desa, the director of NIO, welcomed all the participants on behalf of the host country and the local organizers. Dr Gunnar Kullenberg, the executive secretary of the IOC, addressed the session on behalf of the commission and UNESCO. He emphasized the importance of regional cooperation in marine research, and referred to the programmes being implemented in regional subsidiary bodies of IOC. Subsequently, I addressed the session on behalf of the Government of India, and stressed the importance that our country attached to ocean studies and the development of human resources. I referred to natural hazards and the need to improve forecasting and warning systems at the regional level. Dr S.Z. Qasim, the founder secretary of DOD, delivered the keynote address and recalled various developments in Indian Ocean oceanography.

The executive secretary of IOC invited nominations for the chairman of the session. Dr Zomorrodian, the delegate of the Islamic Republic of Iran, proposed my name. It was seconded by the delegate of Bangladesh, and I was elected with acclamation. I took the chair and invited proposals for the position of vice chairman. The Indian delegate proposed the name of the Islamic Republic of Iran delegate, Dr Zomorrodian. This move was seconded by the delegate of the Maldives, and was accepted with enthusiasm.

I recalled for the benefit of everyone present that the provisional agenda had been circulated from the IOC secretariat and invited comments. Based on the suggestions of the member states, projects related to storm surges, sea level observations, the evaluation of the impacts of oil spills, a specific action programme to host the 1998 International Year of the Ocean, Global Coral Reef Monitoring Network (GCRMN), data directories and data management were proposed to be included in the agenda. After discussions, the agenda for the meeting was finalized with the approval of the members. Dr Sumitra Vijayaraghavan of the NIO informed the session participants about the documentation and reviewed the list of documents, along with the proposed time schedule.

I conducted the session as planned and concluded it with the adoption of the Session Report and Recommendations. The executive secretary of the IOC drew attention to the report of the first session of the committee, summarized the actions taken by the IOCINDIO and noted that the cooperative activities at the regional level had so far been rather limited. He spoke briefly about the developments of the IOC within UNESCO in the preceding four or five years. He suggested that all countries look into the multiple uses of their coastal zones and pursue proactive management, which would ensure a balanced use of the region. The coastal zone had to be maintained as a natural ecosystem. In this context, the exchange and sharing of data were also stressed as a critical need for the region as a whole.

Delegates then presented information on their national marine research programmes, the capacity to undertake marine science activities, their priorities and the areas that might require strengthening. Based on the presentations, I suggested that the IOCINDIO should prepare a synthesis of the information, with explanations on the sufficiency of the data and information. Since the available information was inadequate, it was suggested that such a report be ready for presentation at the 1998 International Year of the Ocean programme. In the course of the session, several topics of interest and relevance to the region were presented, which included the following areas:

1. Capacity-building requirements and technical assistance in a regional cooperative perspective,
2. Programmes of the regional committee for 1997–98/99,
3. Oil pollution and its impact on the Indian Ocean,
4. Regional cooperation to study sea level variations and storm surges,
5. The International Year of the Ocean (YOTO),
6. Coastal zone management,
7. Demarcation of the outer limits of the (legal) continental shelf,
8. Coral reef monitoring and related matters, and
9. Cooperation of the regional committee with other organizations and UNESCO programmes.

The committee re-elected me as the chairman, Dr Zomorrodian as the vice-chairman, and the officers who had served at the second session at Goa for the inter-session period and the third session. In response to my invitation for the proposal on the date and place of the next session of the regional committee, the Islamic Republic of Iran invited the third session of IOCINDIO to be held in their country during November or December 1998. The committee accepted the offer. Further, it adopted the draft summary report and the recommendations. The committee thanked India and its officials for the excellent arrangements. This team included, apart from me, B.N. Krishnamurthy, the director of DOD, Dr E. Desa, the director of NIO, and Dr Sumitra Vijayaraghavan and the staff of NIO, who made the session successful. As scheduled, I closed the session on 22 November 1996.

India's Contributions to the IOC

As a founder member of the IOC, India has been actively associated with its major endeavours such as global and regional ocean observation and capacity-building programmes. Some of the active areas of cooperation proposed for implementation in a phased manner were:

1. Global Ocean Observation System (GOOS),
2. Indian Ocean Global Ocean Observation System (IOGOOS),
3. Data management, data exchange and communication,
4. Ocean services,
5. Capacity building, and
6. Research to resolve issues in operational oceanography in the Indian Ocean, and cooperation among bodies and their programmes.

The International Year of the Ocean (1998)

The IOC initiated the Year of the Ocean (YOTO) proposal, which was subsequently endorsed by the UNESCO General Conference in 1993 and the UN General Assembly. The year 1998 was designated as YOTO concurrently with Portugal's World Exposition in Lisbon, Expo 1998, and celebrations to mark it were proposed. The Expo's theme also centred on the oceans, and a pavilion was to be set up to house UN events and exhibits. It was expected that the IOC would use the opportunity to participate in the Expo in effective ways. In September 1996, the IOC, at its executive council, discussed the manner in which it, its regional bodies, and individual member states might celebrate YOTO.

The IOC's objective was to utilize YOTO for purposes of education and creating public awareness. It was viewed as an opportunity to emphasize the importance of the ocean to all communities including the masses, industries and governments and to promote science, education, culture and so on. Much could be done simply to raise public awareness by including the IOC–YOTO logo on correspondence, meeting arrangements and so on. Conferences and seminars that had already been planned and funded were to be identified as YOTO events at virtually no cost. The IOC suggestions also included the following aspects:

- Developing circumnavigation of the globe by linking together research/training cruises through its various regions as countries came forward to dedicate their ships for these purposes,

- Training opportunities on the cruise legs, and the presentation of a film made of videotapes of the cruises as an education aid or at the Expo 1998, and
- Issue of medals or stamps to honour individuals – educators, scientists, government administrators, sportsmen, artists and musicians – who had dedicated themselves to ocean activities.

India joined the rest of the world in celebrating YOTO. The Union government, state governments, NGOs, public and private institutions, universities, schools and many others enthusiastically participated and made the celebrations memorable. Some of the landmark events include:

(i) Signing of the Ocean Charter,
(ii) Issue of commemorative stamps,
(iii) Special supplements in newspapers to induce public awareness about the importance of oceans, with special articles by experts in the country,
(iv) Issue of postal inland letters with the YOTO theme – 'Ocean is the Common Heritage of Mankind: Protect and Preserve it',
(v) Launch and dedication of Data Buoys under the National Data Buoy Programme (NDBP),
(vi) Distribution of 50,000 copies of 'My Ocean Charter' and 400,000 copies of stickers with the IOC theme to create awareness on the importance of oceans among the youth through educational institutions,

(vii) Beach campaign on 'Save our Beaches and Seas' for one week – to educate the public on the effect of solid waste dumping on coastal and marine ecosystems,

(viii) Hoisting the YOTO flag on ORV *Sagar Kanya* at a public function presided by the Minister for Power and Science and Technology, Government of Goa,

(ix) Programme on-board the ORV *Sagar Kanya* regarding YOTO, and

(x) Forty-seven-day cruise on ORV *Sagar Kanya* from Mormugao Port (Goa) to Port Louis (Mauritius) with twenty-nine scientists (seventeen Indians and twelve foreigners) integrated with the IOC Circumnavigation Cruise of the global sea.

Conclusion

SPACE PROPULSION HAS FASCINATED mankind for years and will continue to do so in the future. India can be proud of its presence among the spacefaring nations that have developed robust and economical space transportation systems. In this context, it would be pertinent to sum up the major kinds of technologies we have employed so far and point to current and emerging trends in the discipline. The Appendix to this work further delineates the technology utilized in space propulsion for orbit raising, orbit manoeuvres, orbit corrections, station acquisition, de-orbiting, attitude correction, de-boosting and so on. It also refers to the future of escape propulsion.

In space propulsion, the chemical propulsion system has attained high levels of maturity and is expected to stagnate in the near future. Chemical propulsion included solid fuel propulsion, liquid propulsion, hybrid propulsion and green propellant propulsion. Of these, developments in escape propulsion system based on liquid oxygen and liquid methane have been gaining significance for future interplanetary space travel and colonization. On the other hand, non-chemical propulsion comprises electric propulsion (ED), tether propulsion, solar sail propulsion, nuclear propulsion, laser propulsion and fusion propulsion.

The electric propulsion system (EPS) is also becoming prominent and the trend is veering towards all-electric-propulsion spacecrafts. New concepts such as tether propulsion and solar sail propulsion are expected to be widely utilized soon. In the area of deep space propulsion, the Radioisotope Thermoelectric Generator (RTG) has been used for many interplanetary missions. In addition, the development towards nuclear propulsion appears to be a promising candidate for deep space propulsion. Thus, the space propulsion arena is becoming more exciting all the time and is expected to enthral mankind for a long time to come.

Nothing ever built rose to touch the skies, unless someone dreamed that it should, believed that it could and willed that it must. The space shuttle programme is an example of extreme innovation and imagination. Built on the pride and commitment of thousands of dedicated men and women, it leaves a legacy highlighted by historic and awe-inspiring achievements in 135 missions over more than thirty years. It has lifted the world's imagination, redefined human space flight and made the international space station an icon for aspiring scientists in the discipline.

We are living in an era of incessant scientific and technological revolution. India's scientific and technological progress in the post-Independence era emerged from the famous 'Homi Bhabha–Vikram Sarabhai' work philosophy of building teams and schemes around individuals whose track record, probity, zeal and concern for social causes are absolutely beyond doubt. They made things happen the way no one else could. ISRO and the Department of Atomic Energy (DAE) are solid manifestations of this vision and continue with the work culture initiated by these two leaders. These organizations boast of a programme-oriented approach, delegations with authority and accountability, monitoring systems, reviews and rewards, and clear-cut documentation. These traits appear very rarely now.

Based on our success over the last six decades, our nation has modified its policy to be in line with the advanced nations of the globe. It calls for redefining our developments in our multi-institutional and multidisciplinary enterprises. The three decades of my association with ISRO convinced me that we can achieve all our goals if we sharpen our professional and managerial skills and put in sincere, dedicated, focused and sustained efforts. To succeed, we should shift our wisdom from words to wisdom in action and set our systems in order.

We should make India an informed society by generating, sharing and applying knowledge. Further, we should promote visionaries and innovators who can persuade millions of people to experiment with technologies that they have never before considered. These visionaries may go against mainstream thinking and follow their intuitions. They not only envisage the technology of the future but also bring their vision to fruition. They should be brave enough to think differently, bold enough to believe that they can, and talented enough to achieve the stuff of dreams.

The significant progress of space technology in ISRO over the past six decades is laudable and globally acknowledged. The triumphant story is due to the contribution of all

ISRO personnel, its leadership, the collaborators and the government. In this book, I have presented the history of the propulsion systems in ISRO in a comprehensive way. I believe that we will do everything necessary to be second to none globally and will ultimately be proud of our achievements.

During my subsequent tenure at DOD, I was able to reorganize the department to be more productive and ambitious in its ventures, to bring in participants from Indian industries and academic institutions into oceanographic research and to initiate foreign collaborations in this discipline.

Today, I am satisfied that the lion's share of my life has been in the service of the nation, as a responsible scientist. I hope that the youth of our country will carry forward the spirit of enquiry in a desirable, proactive manner in all walks of life.

Appendix

IN THIS APPENDIX I have recalled the history of rocket propulsion, classified the propulsion technologies, indicated the boundaries of space propulsion regions and presented in brief different types of advanced propulsion systems being developed and proposed.

A Brief History of Rocket Propulsion

Around 100 BC, a Greek inventor known as Hero of Alexandria came up with a rocket-like device called an aeolipile. His invention depended on the mechanical interaction of heat and water. It used steam for propulsion.

Before Konstantin Tsiolkovsky (1857–1935) could become the famous Russian scientist that he was, he became interested in the possibility of space flight in his teens. He eventually developed the basic theory of rocket propulsion. In 1926, Tsiolkovsky suggested the use of artificial earth satellites, including manned platforms, as way stations for interplanetary flight. In 1929, he put forth the idea of a multistage rocket.

Robert H. Goddard (1882–1945), an American physicist, published several papers on rocket theory. He was awarded patents for a rocket that used solid and liquid propellants, and for a multistage or step rocket. Goddard built and flew the first successful liquid propellant rocket. Of primitive design and construction, the rocket flew across 56 metres at Auburn, Massachusetts, on 16 March 1926. So prolific were his patents that those who followed could hardly take a step without in some way infringing on one or more of them.

Wernher Magnus Maximilian Freiherr von Braun (1912–77) was a German-American aerospace engineer and space architect. He was the leading figure in the development of rocket technology in Germany and a pioneer of rocket technology and space science in the United States. In his twenties and early thirties, von Braun worked in Nazi Germany's rocket development programme. He helped design and develop the V-2 rocket during World War II. Following the war, he was secretly moved to the United States, along with about 1,600 other German scientists, engineers and technicians. He worked for the US Army on an intermediate-range ballistic missile (IRBM) programme, and developed the rockets that launched Explorer 1, the first space satellite of the United States. His group was assimilated into NASA, where he served as director of the newly formed Marshall Space Flight Centre and as the chief architect of the Saturn V superheavy-lift launch vehicle that propelled the Apollo spacecraft to the Moon.

Sergei Pavlovich Korolev (1907–66) worked as the leading Soviet rocket engineer and spacecraft designer during the space race between the United States and the Soviet Union in the 1950s and 1960s. He is regarded by many as the father of practical astronautics. He was involved in the development of the R-7 Rocket and Sputnik 1 and the launch of Laika, the dog; he also successful in launching Yuri Gagarin to outer space, the first human being in outer space.

Space Shuttle

Another great scientific feat in the field of propulsion in the twentieth century is the development of the space shuttle. It was the first operational orbital spacecraft designed for reuse. It carried different payloads to LEO, provided crew rotation and supplies for the International Space Station (ISS), and performed satellite servicing and repair. Each space shuttle was composed of three main assemblies: a reusable Orbital Vehicle (OV), an expendable external tank, and two reusable Solid Rocket Boosters (SRBs). Only the OV entered the orbit shortly after the tank and boosters were jettisoned. The vehicle was launched vertically like a conventional rocket, and the orbiter glided to a horizontal landing like an airplane, after which it was refurbished for reuse. The SRBs parachuted to splashdown in the ocean from where they were towed back to shore and refurbished for later missions.

Classification of Propulsion Technology

Chemical Propulsion

- **Solid Fuel Propulsion:** In solid fuel propulsion, all the components required for combustion are mixed together and packed into a combustion chamber with a nozzle as a single body along with necessary accessories. When lit, the fuel grain will burn energetically, releasing a large volume of hot gases which will be ejected through the nozzle to provide the thrust. It cannot be stopped or restarted.

- **Liquid Propulsion**: Liquid propulsion can be classified into two categories – monopropellant and bipropellant. In a bipropellant system, the propellants exist as two components and are stored in two separate tanks. To initiate combustion and produce thrust, they are pumped into a combustion chamber with a nozzle. This type of propellant is used for a wide range of engines for thrust generation. The next type is monopropellant, in which a single propellant, having the oxidizer bound into the propellant molecule, uses catalysts and burns by itself. Monopropellant systems are used in control systems and in spacecrafts. By regulating the flow of liquid propellants, the liquid propulsion systems can be restarted and shut off and even throttled. Liquid rockets are more complex and have a higher specific impulse than solid propellants.
- **Hybrid Propulsion**: Hybrid propulsion is a combination of solid propulsion and liquid propulsion systems. They are conceived to overcome the complexities of liquid propulsion systems and the lack of controllability of solid propulsion systems. The basic idea is to inject a liquid oxidizer into a solid fuel grain consisting of only fuel that cannot sustain combustion on its own. The engine can be throttled up and down and also shut off and restarted. However, it is difficult to achieve very

large thrusts and, therefore, hybrid propellant engines are rarely built.

- **Green Propulsion:** Green propellants is the name of a family of propellants that satisfy certain requirements such as low toxicity, low pollution, good storability, wide material compatibility, reduced operational hazards and good performance. With increasing level of future activities and applications, green space propulsion becomes significant.

Chemical propulsion systems play an inevitable role in escape propulsion. A large number of solid and liquid propellants are used in these propulsion systems.

Non-Chemical Propulsion

- **Electric Propulsion:** Electric Propulsion (EP) is a class of space propulsion that makes use of electrical power to accelerate a propellant by electrical and/or magnetic means. The use of electrical power enhances the propulsive performances of the EP thrusters, compared to conventional chemical thrusters. The propellant is ejected up to twenty times faster than from a classical chemical thruster and, therefore, the overall system is many times more mass efficient. Unlike chemical systems, electric propulsion requires very little mass for the desired performance of a given spacecraft.
- **Tether Propulsion:** Space tethers are long cables, which can be used for propulsion, momentum exchange, stabilization and attitude control, or to maintain

the relative positions of the components of a large, dispersed satellite/spacecraft sensor system.

- **Solar Sail Propulsion**: Solar sails (also called light sails or photon sails) use radiation pressure exerted by sunlight on large mirrors for propulsion. Solar sail crafts offer the possibility of low-cost operations combined with long operating lifetimes. Since they have few moving parts and use no propellant, they can potentially be used numerous times for delivery of payloads.
- **Nuclear Propulsion:** Nuclear propulsion includes a wide variety of propulsion methods that fulfil the promise of the Atomic Age by using some form of nuclear reaction as their primary power source. Many types of nuclear propulsion have been proposed, and some of them have been tested for spacecraft applications.
- **Laser Propulsion**: Laser propulsion is a form of beam-powered propulsion where the energy source is a remote (usually ground-based) laser system and separate from the reaction mass. This form of propulsion differs from a conventional chemical rocket where both energy and reaction mass come from the solid or liquid propellants carried on-board the vehicle.
- **Fusion Propulsion**: A rocket engine driven by fusion propulsion which could provide efficient and long-term acceleration in space without the need to carry a large fuel supply. For space flight, the main advantage of fusion would be the very high specific impulse, and the main disadvantage the (likely) large mass of the reactor.

Boundaries of Space Propulsion Regions

Based on the regions of operation, beginning from the launch pad on the earth's surface to the end of the heliosphere, space propulsion technologies are classified into the following categories:

(i) Escape Propulsion: It describes propulsion technologies required by space missions where the space vehicle leaves the launch pad on Earth's surface till injected into Earth's orbit.

(ii) In-Space Propulsion: It describes propulsion technologies required by space missions between LEO and geostationary earth orbit. In-Space Propulsion harbours all communication assets, Earth observation systems, navigation, weather and so on.

(iii) Deep-Space Propulsion: It operates beyond Geosynchronous Orbit (GSO) in outer space, encompassing interplanetary, interstellar and intergalactic regions.

Escape Propulsion

Considerable improvements and technological developments have been achieved during the last six decades. The space shuttle is a luxury ship compared to the Mercury capsule that carried the first American astronaut into space. Over the years, the US, Russia, the European Union, China, Japan, India and South Korea have developed a large number of expendable launch vehicles of varying capabilities.

There had been significant developments in the area of escape propulsion over the past three to four decades. Recent trends are towards trading launch vehicle performance efficiency for low cost and high reliability. The major cost-saving avenues involve the use of advanced lightweight materials, simplified and automated manufacturing process, automation in test and inspection procedures, use of modular systems and cheaper propellant.

Liquid oxygen and liquid methane propellant combination is emerging as a promising fuel of the future interplanetary space travel. This is because a spacecraft returning from Mars could extract methane fuel from the Martian atmosphere and use it for the subsequent journey. This made LOX/methane a standard for deep-space manned spacecraft missions. The recent test of the 200 tonne Raptor engine – a full-flow staged-combustion engine that intends to power the next generation of SpaceX launch vehicles for the exploration and colonization of Mars – is an example in this direction. It is noted that many space agencies, including private entrepreneurs, are engaged in active development in this area.

In-Space Propulsion

Spacecraft propulsion is any method used to accelerate spacecraft and artificial satellites. There are many different methods. Each method has drawbacks and advantages, and spacecraft propulsion is an active area of research.

The majority of spacecraft use chemical rockets (bipropellant system or monopropellant system) for producing thrust, which in turn helps in achieving the following functions: orbit rising, orbit manoeuvres, orbit correction, station acquisition, de-orbiting, inclination correction, altitude correction, momentum dumping, and de-boosting.

Chemical Propulsion System

ISRO has a fleet of thrusters developed for various applications. These include the 11N and 1N thruster for the IRS class of satellites and a 200N monopropellant thruster developed for use in scientific missions. In the area of bipropellant thrusters, ISRO has a very reliable engine called the LAM engine, which is used for orbit-raising manoeuvres. It has also developed 22N and 10N thrusters for station-keeping operations. Recently, ISRO has successfully developed an 800N throttleable engine, which is planned to be used in the soft landing of Vikram in Chandrayaan-2 mission. Thus, in the area of chemical propulsion system for spacecrafts, ISRO has gained self-sufficiency in designing and realizing thrusters, and is on par with its counterparts in other spacefaring nations.

Electric Propulsion System

Another emerging area is electric propulsion, which has seen significant increase in usage in the last few decades. Advancements in electric propulsion–related technologies and thruster design improvements have brought some electric propulsion devices to a high level of technological maturity. For select missions, replacing current chemical propulsion systems with high performance electric propulsion systems can provide substantial mass and cost savings, increased orbital lifetimes and increased mission capabilities/flexibility. The major classification of electric thrusters is: (a) electrothermal (b) electromagnetic, and (c) electrostatic.

Electrothermal Propulsion

In electrothermal thrusters, electrical energy is used to heat the propellant, which is expanded through a nozzle. Arc jets, resistojets and microwave arc jets are examples of electrothermal thrusters. The thermal arc jet converts electrical energy into thermal energy by using arc discharge. This thermal energy is converted to kinetic energy using a nozzle for generating thrust. The resistojet uses resistive elements to heat the propellant and thus increases the exhaust velocity. Potential propellants for resistojet include nitrogen, ammonia, hydrogen and carbon dioxide. Microwave arc jets use microwave energy to heat the propellants and the heated propellant is expelled through a nozzle.

Electromagnetic Propulsion

In electromagnetic thrusters, the ions are accelerated by Lorentz force (the force which is excreted by a magnetic field on a moving electric charge). The Magneto Plasma Dynamic (MPD) thruster, Variable Spacecraft Impulse Magneto Plasma Rocket (VASIMR) and pulse plasma thrusters are examples of this type. In the MPD thruster, a gaseous propellant is ionized and fed into an acceleration chamber, where the magnetic and electric fields are created using a power source. The charged particles are propelled by Lorentz force resulting from the interaction between the current flowing through plasma and magnetic field. MPD thrusters have the potential for thrust levels up to 200N, by far the highest for any form of electric propulsion.

A drawback, however, is the requirement of hundreds of kilowatts of power needed for optimum performance.

Current interplanetary spacecrafts have power limitations to meet this high demand. VASIMR bridges the gap between high-thrust, low-specific impulse and low-thrust, high-specific-impulse systems and is capable of functioning in either mode.

The pulse plasma thruster (PPT) works by passing an arc of electricity through the fuel, causing ablation and sublimation of the fuel. The heat generated by this arc causes the resultant gas to turn into plasma thereby creating a charged gas cloud. The electromagnetic field accelerates the plasma at high velocity to generate thrust. PPTs are well suited for relatively small spacecrafts (particularly cubesats) for attitude control, station keeping, etc.

Electrostatic Propulsion

In electrostatic thrusters, the electric field (Coulomb's force) is employed to accelerate the plasma. In a gridded ion thruster, propellant atoms are injected into the discharge chamber and are ionized by electron bombardment, forming plasma. The ions are guided through the extraction holes by the powerful electric field. The final ion energy is determined by the potential of the plasma, which generally is slightly greater than the screen grid voltage. The expelled ions propel the spacecraft in the opposite direction. The ion optics is constantly bombarded by a small amount of secondary ions and erode or wear away, thus reducing engine life.

The stationary plasma thruster (SPT) is a type of electrostatic thrusters in which the propellant is accelerated by an electric field. SPT uses electrostatic potential to accelerate ions to high speed. The ionized propellant is accelerated by the electric field between the anode and cathode. SPTs find application in station keeping and are at times used as the primary propulsion for GTO–GSO orbit raising. All these systems rely on solar power and hence starve for power when the distance between the spacecraft and the sun is very high or in locations where solar energy is available only intermittently.

ISRO has developed an 18mN thruster-based electric propulsion system, which is currently used in the GSAT-9 spacecraft. The next higher version, the 75mN thruster-based electric propulsion system development, is nearing completion. ISRO is also developing 300mN thruster-based electric propulsion system (EPS) for power in an all-electric satellite propulsion system.

Tether Propulsion

Space tethers are long cables which can be used for propulsion, momentum exchange, stabilization and attitude control, or maintaining the relative positions of the components of a large, dispersed satellite/spacecraft sensor system. Depending on the mission objectives and altitude, space flights using this form of spacecraft propulsion are theorized to be significantly less expensive than those using rocket engines. The main techniques for employing space tethers in development are as detailed below.

Electrodynamic Tethers

These are conducting tethers that carry a current that can generate either thrust or drag from a planetary magnetic field, in much the same way as an electric motor does.

Momentum Exchange Tethers

These can be either rotating tethers or non-rotating tethers that capture an arriving spacecraft and then release it at a later time into a different orbit with a different velocity. Momentum exchange tethers can be used for orbital manoeuvring, or as part of a planetary-surface-to-orbit/orbit-to-escape-velocity space transportation system.

Tethered Formation Flying

This is typically a non-conductive tether that accurately maintains a set distance between multiple space vehicles flying in formation.

Solar Sail Propulsion

Solar sails are a proposed method of spacecraft propulsion using radiation pressure exerted by sunlight on large mirrors. Solar sail craft offer the possibility of low-cost operations combined with long operating lifetimes. Since they have few moving parts and use no propellant, they can potentially be used numerous times for delivery of payloads.

Solar sails use a phenomenon that has a proven, measured effect on the spacecraft. Solar pressure affects all spacecrafts, whether in interplanetary space or in orbit around a planet or small body. A typical spacecraft going to Mars, for example, will be displaced thousands of kilometres by solar pressure, so the effects must be accounted for in trajectory planning, which has been done since the time of the earliest interplanetary spacecraft. Solar pressure also affects the orientation of a craft, a factor that must be included in spacecraft design.

Deep-Space Propulsion

The main constraints for the use of conventional propulsion systems for deep-space propulsion are its limitation on the Isp generated (Isp range of a chemical propulsion system is between 400 and 460 seconds), requirement of long period of coasting/use of planet's gravity to steer the vehicle on to its desired trajectory, limited power availability as we move away from Sun (solar power). Three promising propulsion systems which can overcome these drawbacks are:

- Nuclear thermal propulsion,
- Solar thermal propulsion, and
- Use of radioisotope thermoelectric generator (RTG) for powering electric propulsion systems.

Nuclear Thermal Propulsion

A nuclear thermal rocket is a proposed spacecraft propulsion technology. In a nuclear thermal rocket, a working fluid, usually liquid hydrogen, is heated to a high temperature in a nuclear reactor, and then expanded through a rocket nozzle to create thrust. The nuclear reactor's energy replaces the chemical energy of the propellant's reactive chemicals in a chemical rocket. The thermal heater/inert propellant paradigm as opposed to the reactive propellants of chemical rockets turns out to produce a superior effective exhaust velocity, and therefore superior propulsive efficiency, with specific impulses to the order of twice that of chemical engines.

A solar thermal rocket

It is a theoretical spacecraft propulsion system that would make use of solar power to produce heat reaction mass. The rocket would only have to carry the means of capturing solar energy, such as concentrators and mirrors. The heated propellant would be fed through a conventional rocket nozzle to produce thrust. Its engine thrust will be directly related to the surface area of the solar collector and to the local intensity of the solar radiation.

For spacecrafts operating at or near 1 angstrom unit (Å)* from Sun, photovoltaic system has the advantage of cost effectiveness. At greater distance from Sun, radioisotope power systems become favoured options to satisfy the modest power requirement.

Radioisotope Thermoelectric Generator (RTG)

There are lightweight, compact spacecraft power systems that are extraordinarily reliable. RTGs provide electrical power using heat from the natural radioactive decay of plutonium-238, in the form of plutonium dioxide. The heat energy is converted to electric energy through a passive (thermoelectric) or dynamic process (Stirling, Brayton, etc.).

---

* 1 Å -10^{-10} metre

Glossary of Technical Terms

Ablative Systems Division (ABSD): Research and Development Division of Vikram Sarabhai Space Centre (VSSC), responsible for the development of ablative composite components and realization of nozzle assembly for solid rocket motors and throat insert for liquid engines. Ablative liners protect the nozzle hardware of the rockets from extreme thermal environment due to internal flow of hot gases and limits the temperature within acceptable level.

Apogee: When a satellite orbits around a planet in an elliptical orbit, the point at the farthest distance from the planet is called 'apogee' and nearest is called 'perigee'. Both apogee and perigee are the same for a circular orbit.

Argo profiling floats: Argo is an international programme that collects information from inside the ocean using a fleet of robotic instruments that drift with the ocean currents and move up and down between the surface and a mid-water level. Each instrument (float) spends almost all its life below the surface. The name Argo was chosen because the array of floats works in partnership with the Jason earth observing satellites that measure the shape of the ocean surface. (In Greek mythology, Jason sailed on his ship *Argo* in search of the golden fleece). One of Argo's key objectives is to observe ocean data related to climate change. It assesses ocean temperatures, water cycle, and sea level rise. The sensors on Argo floats and the floats themselves are what make the Argo programme possible.

Ariane Passenger Payload Experiment Satellite (APPLE): The Ariane Passenger Payload Experiment Satellite (APPLE) was ISRO's first indigenous, experimental communication satellite with a C-Band transponder. It was launched from Centre Spatial Guyanais, near Kourou in French Guiana, into GTO (Geosynchronous Transfer Orbit) by the third development flight of European Space Agency's Ariane vehicle on 19 June 1981.

Ariane-4: This is one of the family of operational launch vehicles, developed by Ariane Space in Europe as a joint venture of several European countries under the leadership of France.

Attitude and Orbit Control System: This is used in a three-axis-stabilized launch vehicle in flight to maintain the pre-planned attitude and orbit by providing the required forces and moments.

Attitude hold: Attitude hold is holding the attitude. In an aircraft, it is the autopilot setting that will hold the pitch of the aircraft constant when set to this mode.

Augmented Satellite Launch Vehicle (ASLV): ASLV is an all-solid launch vehicle. It is the second 'Technology

Development' Launch Vehicle of ISRO after SLV-3 with the payload capacity to 150 kg, thrice that of SLV-3, for Low Earth Orbits (LEOs). It is configured with the addition of two identical first-stage motors of SLV 3 as zero stage strap-on with no new development in propulsion system.

Booster: For launching satellites with heavier payloads at higher altitudes a multi-rocket system is essential with each stage providing the increment in the velocity and separating out from the launch vehicle. The first stage develops the highest thrust and provides the largest total impulse to the launch vehicle. This first stage is called 'Booster'. Strap-on is also added to provide initial high-thrust capability. The booster can be a solid or liquid propulsion system with solid or liquid strap-on.

Chaff: Chaff is a one-inch-diameter bundle of two-inch-long copper needles of a few micron diameter, ejected by a dart payload of a small-diameter rocket like 'Judi' of the USA at around 50 km altitude. These fine copper needles get disbursed and float as cloud and drift along the wind. This cloud is tracked by a land-based radar to study the wind pattern in lower atmosphere.

Computational Fluid Dynamics (CFD): Computational fluid dynamics is a branch of fluid mechanics that uses numerical analysis and data structures to analyse and solve problems that involve fluid flows. It is a science that, with the help of digital computers, produces quantitative predictions of fluid-flow phenomena based on the conservation laws (conservation of mass, momentum and energy) governing fluid motion.

Convergent–Divergent Nozzle: In rocket propulsion, the forward propulsive thrust is a reaction to the discharge of the high-velocity hot gases of combustion products at supersonic speeds through convergent–divergent nozzles. To generate the high thrust, the products of combustion inside the combustion chamber of the rocket motor has to be accelerated. As well known by the laws of gas dynamics, in subsonic regions of flow the area of the flow passage has to decrease to accelerate the flow up to the throat area where the flow becomes sonic. Beyond that, in the supersonic region, the flow area of the flow passage has to increase to accelerate the flow. Thus convergent–divergent nozzles are used in rocket propulsion to produce thrust to move forward. In these nozzles, the area of flow passage reduces in the convergent region where the flow is sub-sonic up to the throat where it becomes sonic and beyond the throat the area of flow passage increases in the divergent region where the flow is supersonic.

ETA: Explosive Transfer Assembly (ETA) is a confined detonating fuse, consisting of a lead-sheathed mild detonating chord (MDC) covered with polythene sleeve and multilayered fibreglass braiding which is, in turn, protected by external plastic sheath or sleeve. On receipt of detonation stimulus from Initiator/Safe Arm, the confined explosive charge gets ignited and transmits it to the device or system to be actuated/ignited retaining ETA integrity.

Fin Tip Control (FTC)/ Fin Tip Control System: Rockets and some launch vehicles have fins, attached to the base shroud at their rear end for stability in flight. In a Fin Tip Control System, a small portion of the fin at the tip is made movable. During flight, this movable portion of the fins is deflected in desired direction, so that aerodynamic force acting on it steers the rocket/launch vehicle in the specified direction.

Fleet 77: Inmarsat launched Fleet 77, a new maritime satellite service that gave ocean-going vessels access to global voice, data and fax communications. Fleet Safety brings out Global Maritime Distress and Safety System (GMDSS) safety services to tens of thousands of Fleet Broadband and Fleet One users for instant distress alerting, reception of maritime safety information (MSI), voice distress, urgent medical advice and assistance, access to innovative Inmarsat safety services and direct and immediate access to over fifty search and rescue centres around the world.

Flex Nozzle Control (FNC) System: FNC system consists of submerged flex nozzle with contoured or conical divergent and two-actuation systems (electromechanical or electro-hydraulic depending upon the force requirements) at 90 degrees apart with mirror image potentiometers. The actuators are powered by either battery for electromechanical actuators or turbopump/blow down power pack for electro-hydraulic actuators. Flex nozzle consists of flex seal located in the submerged region of nozzle. Flex seal enables deflection (vectoring) of flex nozzle upon application of actuator load on the divergent. Vectoring of the nozzle during the operation of solid motor in flight changes the thrust vector direction depending upon the requirement from autopilot in case of any deviation in the trajectory of the vehicle from the planned flight path.

Geostationary Transfer Orbit (GTO): Satellites are placed normally in geosynchronous orbit in two steps; first it is placed in highly eccentric low earth orbit (LEO) with the apogee close to the final geosynchronous orbit. This orbit is called Geostationary Transfer Orbit. Subsequently the Liquid

Apogee Motor of the satellite is burnt to raise the perigee till the desired geosynchronous orbit is achieved.

Geosynchronous Orbit (GSO): Satellite on Geosynchronous Orbit around a planet has the same angular velocity of the planet. For an observer on the planet the satellite will appear stationary and communication between the satellite and the planet can be continuous all the time.

Geosynchronous Satellite Launch Vehicle (GSLV): Geosynchronous Satellite Launch Vehicle (GSLV) is an expendable launch system operated by ISRO. GSLV was used in fourteen launches from 2001 to 2021, with more launches planned. GSLV is the first ISRO launch vehicle configured with an indigenous cryogenic stage. In late 1990, ISRO established a Cryogenic System Project (CSP) to develop GSLV with a cryogenic stage. Then ISRO entered into a collaboration with Glavkosmos of the then USSR to procure a few cryogenic stages and develop GSLV. The first GSLV was successfully launched from India in April 2001 with a cryogenic stage procured from Russia. In the meantime, ISRO started a cryogenic upper stage (CUS) project in 1995 to develop an upper stage for GSLV indigenously. GSLV with indigenous CUS was launched successfully for the first time in January 2014. Presently GSLV is operational with the indigenous CUS.

Geosynchronous Satellite Launch Vehicle Mark III (GSLV–Mk-III): The Geosynchronous Satellite Launch Vehicle Mark III (GSLV-Mk-III), also referred to as the Launch Vehicle Mark 3 is a three-stage medium-lift launch vehicle developed by ISRO. After the successful use of CUS in GSLV, ISRO developed C25, a cryogenic stage with 28 tonnes of propellant, powered by CE20, a cryogenic engine, developing a 20 tonne thrust for use in the GSLV-Mk-III vehicle. This vehicle has a diameter of 4 metres lift mass of 640 tonnes, heat shield diameter of 5 metres and a height of 43.43 metres. The first development flight of the GSLV-Mk-III-D1, using this stage, was a grand success in its maiden attempt in July 2017. GSLV-Mk-III is an operational vehicle, to be deployed for Chandrayaan-2, Chandrayaan-3, and Gaganyaan missions.

Global Positioning System (GPS): The Global Positioning System (GPS) is a satellite-based radio-navigation system owned by the United States government. It is one of the global navigation satellite systems (GNSSs) that provides geographic location and time information to a GPS receiver anywhere on or near Earth.

HTPB: ISRO has developed composite solid propellants using different types of resins, viz., HTPB (Hydroxyl-Terminated Polybutadiene), PBAN (Polybutadiene Acrylonitrile), CTPB (Carboxyl-Terminated Polybutadiene), etc. Resin acts as a binder in the composite propellant. In ISRO, after the initial development with polyester, PBAN and CTPB resins, HTPB-based formulation has been established and became the workhorse formulation for all solid motors including the largest motor S200 employed in GSLV Mk-III. Major ingredients of the HTPB-based solid propellant are HTPB resin as a binder (14 per cent), ammonium perchlorate (68 per cent) as oxidizer, aluminium powder (18 per cent) as fuel along with curator and other process aids in smaller quantities. Propellant formulation with HTPB resin has better mechanical and ballistic properties.

Indian National Satellite System (INSAT): The Indian National Satellite System or INSAT, is a series of multipurpose geostationary satellites launched by ISRO to satisfy the telecommunications, broadcasting, meteorology, and search and rescue operations. Commissioned in 1983, INSAT is the largest domestic communication system in the Indo-Pacific Region.

Indian Regional Navigation Satellite System (IRNSS): The Indian Regional Navigation Satellite System (IRNSS), with an operational name of NavIC (acronym for NAVigation with Indian Constellation; also, *navik*, meaning 'sailor' or 'navigator' in Indian languages) is an autonomous regional satellite navigation system that provides accurate real-time positioning and timing services.

Indian Remote Sensing (IRS) programme: India's remote sensing programme was developed with the idea of applying space technologies for the benefit of humankind and the development of the country. The programme involved the development of three principal capabilities. The first was to design, build and launch satellites to a sun-synchronous orbit. The second was to establish and operate ground stations for spacecraft control and data transfer along with data processing and archival. The third was to use the data obtained for various applications on the ground.

Inertial Upper Stage (IUS): The Inertial Upper Stage (IUS), originally designated the Interim Upper Stage, by the United States Air Force beginning in 1976, was for raising payloads from Low Earth Orbit to higher orbits or interplanetary trajectories following launch aboard a Titan 34D or Titan IV rocket as its upper stage, or from the payload bay of the space shuttle as a space tug.

Boeing was the primary contractor for the IUS while Chemical Systems Division of United Technologies built the

IUS solid rocket motors. The first launch of the IUS was in 1982 on a Titan 34D rocket from the Cape Canaveral Air Force Station shortly before the STS-6 space shuttle mission.

Interference Fit: When two components are assembled, the dimensions of the mating parts control the fit. Depending upon the functional requirements, relative movements are permitted in some assemblies and are called loose fits. Whereas in some other components relative movements between parts are not permitted and the two components have to function as a single one. The term 'interference fit' is used in such cases. In an interference fit the diameter of the hole will be smaller than the diameter of the mating shaft. Heating of the hole or cooling of the shaft will be done to facilitate the assembly of the interference fit. Depending on the interference required, the tolerance of the components will be specified. In brief, an interference fit, also known as a press fit or friction fit, is a form of fastening between two tight-fitting mating parts that produces a joint which is held together by friction after the parts are pushed together.

Liquid Apogee Motor (LAM): LAM is a liquid motor developing a thrust of around 400N and capable of burning for thousands of seconds continuously. It also has multi-burn, on–off capability. It is also used to transfer the satellite from one orbit to a different one around the same planet and also from one planet to another. It is used to raise a satellite from GTO to GSO in a single or multiple burnings.

Mission Control Facility (MCF): A mission control centre is a facility that manages space flights, usually from the point of launch until landing or the end of the mission. Master Control Facilities at Hassan in Karnataka (and Bhopal in Madhya Pradesh) monitor and control all the geostationary/

geosynchronous satellites of ISRO, namely, INSAT, GSAT, Kalpana and IRNSS series of satellites. MCF is responsible for orbit raising of satellites, in-orbit payload testing, and on-orbit operations all through the life of these satellites. MCF activities include round-the-clock tracking, telemetry and command (TT&C) operations, and special operations like eclipse management, station-keeping manoeuvres and recovery actions in case of contingencies. MCF interacts with user agencies for effective utilization of satellite payloads and to minimize service disturbances during special operations.

Modular Ocean Model (MOM): The Modular Ocean Model (MOM) is a numerical representation of the ocean fluid with applications from the process scale to the planetary circulation scale. It is a three-dimensional ocean circulation model, designed primarily for studying the ocean climate system. INCOIS has setup a high resolution regional Indian Ocean model with open boundary condition in collaboration with National Oceanic and Atmospheric Administration (NOAA)/Geophysical Fluid Dynamics Laboratory (GFDL), Princeton, USA, and Indian Institute of Science (IISc) Bengaluru, using the recent version of the Modular Ocean Model (MOM).

Momentum dumping: Momentum wheels are used in spacecraft attitude control system to stabilize the desired attitude of the spacecraft consuming renewable electrical energy. The disturbance torques on the spacecraft increase the angular momentum of the spacecraft. This change results in speeding of the momentum wheels. The angular momentum stored in the momentum wheels should be removed by implementing external torques to prevent saturation of the momentum wheel motors. When the reaction/momentum wheels are saturated, the magnetic torquers or torue rods are used to remove some of the momentum. The momentum dumping is done by the magnetic torque rods.

Moon Impact Probe (MIP): The Moon Impact Probe developed by ISRO was a lunar probe that was released by ISRO's Chandrayaan-1 lunar remote sensing orbiter which in turn was launched on 22 October 2008 aboard a modified version of ISRO's Polar Satellite Launch Vehicle. It discovered the presence of water on Moon.

The MIP separated from the Moon-orbiting Chandrayaan-1 on 14 November 2008 at 8.06 p.m. and crashed, as planned, out of the lunar south pole after a controlled descent. The MIP struck the Shackleton Crater at 8.31 p.m. on 14 November 2008 releasing underground debris that could be analysed by the orbiter for presence of water/ice. With this mission, India became the fourth nation to reach the lunar surface. Other countries to have done so are the former Soviet Union, the United States and Japan.

Multistage Rocket (MSR): This is a development project of VSSC under Rohini series for studying and gaining experience in some of the technologies involved in developing launch vehicles, like staging, clustering, etc.

Non-destructive Testing (NDT): This is a methodology used to test any article/component without causing any physical damage and affecting its characteristics.

Payload: It includes equipment, sensors and instrumentation systems carried by sounding rockets inside the nose cone and satellites inside its structure for carrying out observations and performing the intended functions to achieve the objectives of the mission.

Perigee: When a satellite orbits around a planet in an elliptical orbit, the point at the farthest distance from the planet is called 'apogee' and that in the rest is called 'perigee'. Both apogee and perigee are the same for a circular orbit.

Pitch control: A vehicle in flight has six degrees of freedom – plus or minus in pitch, yaw and roll directions. To steer the vehicle in the pre-planned orbit, corrective actions have to be taken when deviations are detected by the guidance system. Pitch control system provides the corrective pitch control forces to steer the vehicle along the pre-planned orbit.

Polar Satellite Launch Vehicle (PSLV): Polar Satellite Launch Vehicle (PSLV) is the third-generation launch vehicle of India. It is the first Indian launch vehicle to be equipped with liquid stages. During 1994–2017, the vehicle has launched forty-eight Indian satellites and 209 satellites for customers from abroad. It is a four-stage, expendable, medium-lift launch vehicle with the first and third stages powered by solid propulsion, and the second and fourth stages powered by liquid propulsion. It is designed and operated by ISRO. The organization built five variants of the PSLV based on payload capacities, varying from 600 kg in Low Earth Orbit (LEO) to 1,900 kg in a Sun-Synchronous Orbit (SSO).

Pressure Transducer: It is an electromechanical device to measure the pressure in a specified system. It has a sensing element on which strain gauges are mounted. When the sensing element is exposed to the pressure to be measured, it gets strained and the strain is measured by the strain gauges. These are recorded and translated to the pressure using the established relationship by pressure calibration.

Proof pressure: In space programmes, all components and systems which are exposed to pressure during operations are cleared for use after pressure tests. The level of the pressure to be tested is higher than the operating pressure. The margin over the maximum operating pressure depends on the functions and the development philosophy. Normally for control components it will be 50 per cent higher and for metallic, solid rocket cases it will be 10 per higher.

Qualification (of an engine): The engine is a system with a number of subsystems and each one of them having a number of components. Every one of them have to perform as designed. In a development project, first the design is carried out and it is reviewed and approved. The design will comply with the basic laws of thermodynamics and conservation principles of mass, momentum and energy. Since our systems do not operate on ground, we have no access to take any corrective actions in the event of any malfunction during operation. So every component, subsystem and the integrated system will have to be tested and qualified during the development phase, and design margins will have to be established. Further, they will have to be acceptance-tested as per quality protocol before use. The quality assurance protocol has to be evolved and reviewed by a group of experts and accepted. Tests at component, subsystem and system levels will have to be conducted as per the quality assurance protocol and delivered for flight. Though the system has successfully passed through the qualification test with higher design margins, they will have to pass through the acceptance tests before use in every flight.

Reaction Control System (RCS): Reaction Control Systems are mounted in the base shroud of the stages of the launch vehicle. They use bipropellant or monopropellant liquid engines with on–off capability. Its specifications depend on the launch vehicle's mission requirements. The bipropellant system uses hypergolic propellants which initiate combustion when coming into contact with each other while the monopropellant system requires a catalyst to initiate combustion/ decomposition. The RCS uses pressure-fed feed system which has a high-pressure gas (nitrogen or helium) stored in a high-pressure bottle. A pressure regulator is used to reduce the pressure to the designed value and pressurize the propellant tanks where propellants are stored. This propellant is fed to the engine through a solenoid valve which is commanded to open to permit the flow of the propellant, or to close to stop the flow. When the flow of propellant is permitted, combustion takes place and the discharge of the products of combustion through the nozzle generates the thrust/force. The thrust/force is shut down when the flow is stopped. Thus the liquid engine produces thrust when required by the guidance system to steer the vehicle in the desired flight path. The RCS provides forces in the pitch, yaw and roll directions of the vehicle.

Reusable Launch Vehicle Technology Demonstrator (RLV TD): Reusable Launch Vehicle Technology Demonstration Programme is a series of technology demonstration missions that have been considered as a first step towards realizing a two-stage-to-orbit (TSTO), fully reusable vehicle.

Roll control: A vehicle in flight has six degrees of freedom – plus or minus in all three directions: pitch, yaw and roll. To steer the vehicle in the pre-planned orbit, corrective actions have to be taken when deviations are detected by the guidance system. Roll control system provides the corrective roll control to steer the vehicle along the pre-planned orbit.

Satellite Launch Vehicle-3 (SLV-3): Satellite Launch Vehicle-3 was India's first experimental satellite launch vehicle, which was an all solid, four-stage vehicle weighing 17 tonnes with a height of 22 m and capable of placing 40 kg class payloads in Low Earth Orbit (LEO). SLV-3 was successfully launched on 18 July 1980 from Sriharikota Range (SHAR), when Rohini satellite, RS-1, was placed in orbit, thereby making India the sixth member of an exclusive club of spacefaring nations. SLV-3 employed an open loop guidance (with stored pitch programme) to steer the vehicle in flight along a predetermined trajectory. The first experimental flight of SLV-3, in August 1979, was only partially successful. Apart from the July 1980 launch, there were two more launches held in May 1981 and April 1983, orbiting Rohini satellites carrying remote sensing sensors.

Secondary Injection Thrust Vector Control (SITVC): In the SITVC system, a liquid is injected from the designed location of the nozzle divergence into the supersonic flow of the hot products of combustion. Consequently, there is a shock chemical reaction and flow separation which causes the deflection of the engine thrust. The thrust deflection is a function of the flow rate, pressure and chemical composition of the fluid injected and the location of injector in the nozzle. This deflection of the thrust vector provides the required corrective forces in the pitch and yaw directions.

Specific Impulse (Isp): Isp is the thrust produced by unit weight flow rate of the propellant through the nozzle of the engine and is measured in seconds.

Static test: In launch vehicle missions, the rocket propulsion systems, both solid and liquid, operate in space. There is no scope for any corrective measures if the system malfunctions. Also there is no scope to instrument the system to measure its performance. Hence special test stands are designed, established and mounted on to test beds on ground. The propulsion systems are mounted on to the test bed and held to the ground. They are instrumented fully to measure thrust, pressure, temperatures, vibrations, strains, etc., as per the test plan and connected to the host of digital and analog recording systems. Adequate safety measures are taken and the system is tested in static condition for its full operation. While solid motors are burnt for the full duration, the liquid engines are tested for short durations and in pulse mode, in addition to the full duration and at times beyond too. These static tests provide adequate inputs to estimate the performance of propulsion systems in flight with reliable inputs.

Station keeping: After successful launch, the satellite is to be placed in space at the required orbit and orientation. During the life of its operation, due to disturbances in space and degradation of performance of some of the on-board subsystems, the satellite could shift from its designated position and orientation. Remedial actions need to be taken by the on-board RCS to make corrections in the plane and altitude of the orbit. These operations of the on-board RCS based on ground commands is called station keeping.

Storm surge prediction model: Storm surge models are useful tools for predicting real-time storm surge and for future planning. These models provide real-time forecasting of the water level and potential inundations under storm conditions. Storm surge models are also used to hind-cast the flooding generated by past hurricanes, assess future flood risks and produce potential flooding maps for planning purposes. Though there are many models, they all share the common set of data required for storm surge simulations. A basic recipe for running a storm surge model includes the computational mesh that covers the area of interest, bathymetry data of the area, hurricane information, and atmospheric pressure and tidal information. Most storm surge models are two-dimensional, meaning that the current variation with depth is not resolved in these models. Researchers have shown that a well-tuned, two-dimensional model is sufficient to produce accurate results. However, for some applications such as pollutant transport, a three-dimensional data is necessary for accurate results.

Sustainer: In a two-stage rocket, the bottom stage which ignites first is called Booster Stage. The second stage which separates from the booster and continues with the payload is called the Sustainer.

System Reliability (SR): System reliability relates to ensuring specified performance under specified conditions. The activities involved are many, including, specification review, design review, establishing protocols for qualification testing and acceptance testing for all components, subsystems and integrated system, theoretical studies, modelling, pilot-scale experimentation, verification of material characteristics, manufacturing processes, etc. Further, inspection, quality surveillance, quality control, performance analysis, failure mode evaluation, criticality analysis, failure analysis, corrective measures, estimation of reliability based on performance data are also involved. In addition, test and evaluation and certification of equipment, machinery and test rigs are also part of the activities of System Reliability.

TBI: Through bulkhead initiator (TBI) has a metallic barrier, 2 to 2.5 mm thick, which physically separates the externally located donor charge and acceptor charge in the ignition train. This physical separation eliminates any leakage from the ignition train. During operation, the detonation shock wave is transmitted from donor charge through a solid metal interface to an acceptor charge which will initiate the ignition of intermediate charge or booster charge in the igniter which, in turn, will ignite the igniter grain. The igniter will give the required heat energy to the propellant grain in the main motor resulting in its ignition. TBI is assembled on the igniter head end. PETN (pentaerithritol tetranitrate) explosive charge is used as both donor and acceptor charge, and it has the ability to transfer the shock through the metallic barrier.

Thermal Protection System (TPS): During the operation of the solid motors, combustion products at very high temperature flow through the propellant grain and are discharged at high speed through the nozzle. The motor case, which houses the propellant grain, is exposed to hot gases towards the end of burning, and the nozzle through which the hot gases flow at high speeds experiences high heat flux during the full operation of the motor. Thus the heat transfer phenomenon to the hardware is different in both cases. So far as the motor cases are concerned, a good insulator can provide the thermal protection, and in the case of the nozzle, an ablative material with mass transfer cooling process is required. ISRO has developed both insulating materials and ablating materials as the thermal protection system (TPS). Nitrile-based or EPDM-based formulation is used as thermal protection system for motor cases. Ablative materials like carbon/phenolic (CP) and silica/phenolic (SP) are used as thermal protection system for nozzle hardware. The throat region of the nozzle where the heat flux is maximum use graphite with CP and SP backup.

Thrust Vector Control System: Thrust vector control is achieved by injecting a liquid into the divergent part of the nozzle as in the case of SITVC or deflecting the divergent of the nozzle and changing the direction of the thrust as in flex nozzle control system.

Thrust vectoring: Thrust vectoring, also known as thrust vector control (TVC), is the ability of an aircraft, rocket, or other vehicle to manipulate the direction of the thrust from its engine(s) or motor(s) to control the attitude or angular velocity of the vehicle. In missile literature originating from Russian sources, thrust vectoring is often referred to as gas-dynamic steering or gas-dynamic control.

Yaw control: A vehicle in flight has six degrees of freedom – plus or minus in pitch, yaw, and roll directions. To steer the vehicle in the pre-planned orbit, corrective actions have to be taken when deviations are detected by the guidance system. Yaw control system provides the corrective yaw control forces to steer the vehicle along the pre-planned orbit.

Acknowledgements

THIS BOOK PRESENTS THE history of technical developments during my journey through the Space and Ocean programmes of our nation. All through the book, I have acknowledged the sincere and dedicated contributions of all my colleagues for working as a team under my leadership. I have great pleasure in writing this book, though late, at the request of some of my colleagues and well-wishers as a comprehensive document for record and reference by the future researchers, historians and journalists.

I thank and acknowledge my former colleague Mr Y.S. Rajan for motivating and encouraging me to write this book and for his suggestions regarding its structure and shape. While writing this book, several colleagues of mine gave valuable inputs which I acknowledge with sincere thanks.

I also thank my former colleague Dr Varadarajan Sampath, Department of Ocean Development, presently, Ministry of Earth Sciences, Government of India, for his assistance in final compilation of the photos for the Ocean sector, Mr Eswaran V. Gounder of the PSLV project for final compilation of the photos in the solid propulsion sector and Mr Robin Alexander of LPSC for his assistance for printing photos from my power point documents.

I wrote the first draft of the manuscript based on my lectures, power point presentations and technical reports over the years. Ms Bindu Radhakrishnan edited and prepared it for publication as a book. I acknowledge her valuable contributions.

Last but not the least, I acknowledge and thank senior editor Antony Thomas, executive publisher Udayan Mitra and the management of HarperCollins for publishing this book.

Index

About the Author

Dr Abraham E. Muthunayagam, a mechanical engineer with a PhD from Purdue University, USA, joined INCOSPAR, the predecessor of ISRO, in 1966, where he contributed extensively to the development of propulsion technologies as the founder–director of the organization's Liquid Propulsion Systems Centre.

From 1996 to 2001, he was secretary, Department of Ocean Development, Government of India; and chairman, Regional Committee of Intergovernmental Oceanographic Commission for the Central Indian Ocean. He was vice-chairman, Intergovernmental Oceanographic Commission (1996–98) and chairman, Commission for the Conservation of Antarctic Marine Living Resources (1998–2000).

He has been the founder–vice chancellor of Karunya University, Coimbatore (2001–04); executive vice president, Kerala State Council for Science and Technology and Environment (2005–07); chairman, Board of Governors, IIT Madras (2005–08); and chairman, State Environment Impact Assessment Authority, Kerala (2011–13).

Dr Muthunayagam has received several national accolades, including the Dr V.M. Ghatage Award from the Aeronautical Society of India (1989), the Mechanical Engineering Design Award (1995) from the Institution of Engineers (India), the Life Time Achievement Award in Ocean Science and Technology (2007) from the Ministry of Earth Sciences, the ISRO Outstanding Achievement Award (2008) and the Aryabhata Award (2010) for his contribution in the area of propulsion technology.